关于核能未来的联合建议

——中法三院核能合作（第一期）

Joint Recommendations for the Nuclear Energy Future

A Collaborative Programme between Chinese and French Academies（Phase I）

中国工程院（Chinese Academy of Engineering）

法国国家技术院（National Academy of Technologies of France）　编著

法国科学院（French Academy of Sciences）

科学出版社

北京

内 容 简 介

本书基于中法三院（中国工程院、法国国家技术院、法国科学院）对核能领域的联合研究成果，详细分析了核能发展进程、安全性、面临的挑战及可能的解决措施，内容涉及放射性废物管理、先进核能系统的开发与部署、新项目融资及投资控制，以及公众对核电安全日益提高的要求等，对核能未来发展提出了意见和建议，将有效促进核能的健康发展和进一步提升中法两国对世界核能技术发展的贡献。

本书主要面向关注核能发展的科研工作者、管理人员和有志投身核能事业的青年学子，力图立足科学立场阐释公众关心的问题，对于提高国际核能领域科学技术界的交流水平、增强在核能和平利用上的共识具有重要意义。

图书在版编目(CIP)数据

关于核能未来的联合建议：中法三院核能合作：第一期／中国工程院，法国国家技术院，法国科学院编著 .—北京：科学出版社，2023.9
ISBN 978-7-03-076480-5

Ⅰ.①关…　Ⅱ.①中…②法…③法…　Ⅲ.①核技术-国际科技合作-中、法　Ⅳ.①TL

中国国家版本馆 CIP 数据核字（2023）第 185325 号

责任编辑：周　涵／责任校对：彭珍珍
责任印制：张　伟／封面设计：无极书装

科学出版社 出版
北京东黄城根北街 16 号
邮政编码：100717
http://www.sciencep.com

北京中科印刷有限公司 印刷
科学出版社发行　各地新华书店经销
*
2023 年 9 月第 一 版　开本：787×1092　16
2023 年 9 月第一次印刷　印张：13 1/2
字数：240 000

定价：148.00 元
（如有印装质量问题，我社负责调换）

About the book

The book is a formal publication of a report titled *Joint Recommendations for the Nuclear Energy Future* which was released in 2017 by the Chinese Academy of Engineering, the National Academy of Technologies of France and the French Academy of Sciences. The authors review history, safety, challenges and potential solutions of nuclear development, and present a comprehensive analysis of many aspects related to nuclear energy, including management of radioactive waste, development and deployment of advanced nuclear energy, financing of new nuclear projects and control of costs, and increasing demand of the public for nuclear safety. The authors also put forward proposals and recommendations for the future of nuclear energy, which will effectively promote healthy development of nuclear energy and will add to China and France contributions to the world nuclear energy.

The main target audiences of the book are scientists, researchers and managers who are engaging in nuclear power, and young students who plan to pursue careers in the area. The authors take a science-based perspective to respond to the public concerns about nuclear energy. The book is expected to play a part in facilitating academic exchanges in international nuclear communities and enhancing consensus on peaceful uses of nuclear energy.

作 者 简 介

作者介绍可在法国国家技术院、法国科学院、中国工程院的网站上找到。以下只
列出作者的基本任职信息。

法国国家技术院

http://www.academie-technologies.fr/en/members

Alain BUGAT（项目联合组长）法国国家技术院荣誉院长、院士。曾任法国原子
能署主席，目前是 NUCADVISOR 咨询公司的副总裁。

Yves BAMBERGER 法国国家技术院院士，法国电力公司研发部前主任。

Pascal COLOMBANI 法国国家技术院院士。曾任法国原子能署行政主管、阿海珐
集团监督委员会主席，目前为 A.T Kerney（巴黎）顾问委员会主席。

Bernard ESTEVE 法国国家技术院院士，道达尔公司前核工业顾问，目前任 B.E.
咨询公司董事长。

Gerard GRUNBLATT 法国国家技术院院士，阿尔斯通公司超导应用前经理。

Patrick LEDERMANN 法国国家技术院院士，阿尔斯通印度有限公司前董事总经理。

Philippe PRADEL 法国国家技术院院士，法国 Engie 集团副总裁。

Bruno REVELLIN-FALCOZ（国际协调员）法国国家技术院荣誉院长、院士，达
索航空公司前副总裁。

Bernard TARDIEU 法国国家技术院院士，科因贝利公司荣誉总裁。

Dominique VIGNON 法国国家技术院院士，法玛通公司前总裁兼总经理。目前是
NUCADVISOR 咨询公司合伙人。

法国科学院

http://www.academie-sciences.fr/en/Members/members-of-the-academie-des-sciences. html

Sébastien CANDEL（项目联合组长）法国科学院院长，工程科学专家，中央理工-高等电力学院（萨克雷大学）荣誉教授，近期新任法国电力公司科学委员会主席。

Edouard BREZIN 法国科学院荣誉院长、院士，统计物理和粒子物理专家，巴黎高等师范学校荣誉教授。

Robert GUILLAUMONT 法国科学院院士，放射化学专家，奥赛大学荣誉教授，国家评估委员会成员。

中国工程院

http://en. cae. cn/en/Member/Member/

赵宪庚（项目联合组长）中国工程院院士，中国工程院原副院长。

叶奇蓁（项目副组长）中国工程院院士，核反应堆、核电技术专家，曾任秦山核电站一期总工程师。

中国工作组

中国工作组成员来自中国核工业集团有限公司、中国核科技信息与经济研究院、中国核动力研究设计院、中国核电工程有限公司、中国中原对外工程有限公司、中国辐射防护研究院、中国原子能科学研究院、中广核集团公司研究中心。

田佳树（中国核工业集团公司）（工作组组长），张萌（中国核科技信息与经济研究院），陈彬（中国核动力研究设计院），郭昊（中国核电工程有限公司），邓伟（中国核电工程有限公司），谢小钦（中国中原对外工程有限公司），李幼忱（中国辐射防护研究院），任晓娜（中国辐射防护研究院），尹向勇（中广核集团公司研究中心）

技术与支持组

Wolf GEHRISCH（法国国家技术院）（技术秘书），Jean-Yves CHAPRON（法国科学院）（技术秘书），田琦（中国工程院）（国际协调员），宗玉生（中国工程院），李若瑜（中国原子能科学研究院）（项目联络人）

致谢

全体作者对以下人士仔细阅读报告初版并提供了诸多有用建议表示感谢：Antoine DANCHIN（法国科学院），Jean FRÊNE（法国国家技术院），Ghislain de MARSILY（法国科学院、法国国家技术院），Marc FONTECAVE（法国科学院），Yves LÉVI（法国国家技术院），Olivier PIRONNEAU（法国科学院）。

全体作者也对以下人士所提供的宝贵意见和建议表示感谢：杜祥琬（中国工程院），王大中（中国科学院），郑健超（中国工程院），潘自强（中国工程院），李冠兴（中国工程院），陈念念（中国工程院），于俊崇（中国工程院），徐銤（中国工程院），孙玉发（中国工程院），万元熙（中国工程院），彭先觉（中国工程院），张华祝（中国核能行业协会），雷增光（中国核工业集团公司），白云生（中国核科技信息与经济研究院），石磊（中国核科技信息与经济研究院），刘一哲（中国原子能科学研究院），董玉杰（清华大学），李翔（中国核动力研究设计院），秦忠（中国核动力研究设计院），高瑞发（中国核电工程有限公司），王旭宏（中国核电工程有限公司），杨球玉（中国核电工程有限公司），吕涛（中国核电工程有限公司），李星宇（中国核电工程有限公司），赵树峰（中国核电工程有限公司），方浩宇（中国核动力研究设计院），杨勇（中国原子能科学研究院），陈公全（中广核集团公司研究中心），郑保军（中国核电工程有限公司），马超（中国核电工程有限公司），林浩森（中国中原对外工程有限公司），邵成懂（中国中原对外工程有限公司），任越（中国辐射防护研究院），王彦（中国辐射防护研究院），汪志宇（江苏核电有限公司），刘仲华（中国核工业集团公司），吴国凯（中国工程院），宋德雄（中国工程院），王振海（中国工程院）。

About the Authors

Information concerning the authors curriculum vitae may be obtained from the web sites of the National Academy of Technologies of France, French Academy of Sciences and Chinese Academy of Engineering. Short biographical data are given below.

National Academy of Technologies of France (AT)

http://www. academie-technologies. fr/en/members

Alain BUGAT (Study co-leader) is Member and Honorary President of the National Academy of Technologies of France. He is Former Head of the Commissariat à l'Energie Atomique (CEA) and is currently Vice-President of NUCADVISOR.

Yves BAMBERGER is Member of the National Academy of Technologies of France and Former Director of Research and Development at EDF.

Pascal COLOMBANI is Member of the National Academy of Technologies of France. He is Former General Administrator of CEA, Former President of the "Conseil de surveillance" of AREVA, and is chairing the Advisory Board of A. T. Kerney Paris.

Bernard ESTEVE is Member of the National Academy of Technologies of France and Former Nuclear Counsellor for Total. He is currently President of B. E. Consult.

Gerard GRUNBLATT is Member of the National Academy of Technologies of France. He is Former Head of superconductivity applications at ALSTOM.

Patrick LEDERMANN is Member of the National Academy of Technologies of France and Former Managing Director of ALSTOM India limited.

Philippe PRADEL is Member of the National Academy of Technologies of France and Vice-President of ENGIE Nucléaire, France.

Bruno REVELLIN-FALCOZ (International coordinator) is Member and Honorary President of the National Academy of Technologies of France. He is Former Vice-President Director-General of Dassault Aviation.

Bernard TARDIEU is Member of the National Academy of Technologies of France and Honorary President of COYNE and BELLIER.

Dominique VIGNON is Member of the National Academy of Technologies of France and Former President Director-General of Framatome. He is currently partner of NUCADVISOR.

French Academy of Sciences (AS)

http://www. academie-sciences. fr/en/Members/members-of-the-academie-des-sciences. html

Sébastien CANDEL (Study co-leader) is President of the French Academy of Sciences. He is a specialist in Engineering Sciences, University Professor Emeritus at Centrale Supélec, University Paris-Saclay. He has recently been appointed Chairman of the Scientific council of EDF.

Edouard BREZIN is Member and Past President of the French Academy of Sciences. He is a specialist in Statistical and Particle Physics and Professor Emeritus at Ecole Normale Supérieure.

Robert GUILLAUMONT is Member of the French Academy of Sciences. He is a specialist of Radiochemistry, Honorary Professor at the University of Orsay and Member of the Commission Nationale d'Evaluation.

Chinese Academy of Engineering (CAE)

http://en. cae. cn/en/Member/Member/

ZHAO Xiangeng (Study co-leader) is Member and Past Vice President of CAE.

YE Qizhen (Assistant co-leader) is Member of the CAE. He is a specialist in the field of Nuclear Reactor and Nuclear Power Generation Technology. He was Chief Design Engineer of the Qinshan Nuclear Power Project.

Chinese Working Team

Members of working team are from CNNC (China National Nuclear Corporation), CINIE (China Institute of Nuclear Information and Economics) , NPIC (Nuclear Power Institute of China), CNPE (China Nuclear Power Engineering Corporation), CNOS (China National Nuclear Corporation Overseas Ltd.), CIRP (China Institute for Radiation Protection), CGN (China General Nuclear Power Corporation).

TIAN Jiashu (CNNC) (Team leader), ZHANG Meng (CINIE), CHEN Bin (NPIC), GUO Hao (CNPE), DENG Wei (CNPE), XIE Xiaoqin (CNOS), LI Youchen (CIRP), REN Xiaona (CIRP), YIN Xiangyong (CNG)

Technical and Support Team

Wolf GEHRISCH (AT) (Technical secretary), Jean-Yves CHAPRON (AS) (Technical secretary), TIAN Qi (CAE) (International coordinator), ZONG Yusheng (CAE), LI Ruoyu (CIAE) (Communication)

Acknowledgments

The authors wish to thank Antoine DANCHIN (AS), Jean FRÊNE (AT), Ghislain de MARSILY (AS and AT), Marc FONTECAVE (AS), Yves LÉVI (AT), Olivier PIRONNEAU (AS) for their careful reading of the initial version of this report and for their many helpful comments.

The authors would also like to express their gratitude to DU Xiangwan (CAE), WANG Dazhong (CAS), ZHENG Jianchao (CAE), PAN Ziqiang (CAE), LI Guanxing (CAE), CHEN Niannian (CAE), YU Junchong (CAE), XU Mi (CAE), SUN Yufa (CAE), WAN Yuanxi (CAE), PENG Xianjue (CAE), ZHANG Huazhu (CNEA), LEI Zengguang (CNNC), BAI Yunshen (CINIE), SHI Lei (CINIE), LIU Yizhe (CIAE), DONG Yujie (TSU), LI Xiang (NPIC), QIN Zhong (NPIC), GAO Ruifa (CNPE), WANG Yuhong (CNPE), YANG Qiuyu (CNPE), LV Tao (CNPE), LI Xingyu (CNPE), ZHAO Shufen (CNPE), FANG Haoyu (NPIC),

YANG Yong (CIAE), CHEN Gongquan (CGN), ZHENG Baojun (CNPE), MA Chao (CNPE), LIN Haomiao (CNOS), SHAO Chendong (CNOS), REN Yue (CIRP), WANG Yan (CIRP), WANG Zhiyu (JNPC), LIU Zhonghua (CNNC), WU Guokai (CAE), SONG Dexiong (CAE), and WANG Zhenhai (CAE) for providing valuable comments and remarks.

前　　言

2016 年下半年，中法三院共同组建了一个中法联合研究小组，其总体目标是起草一份概括相关政策和技术方案（包括安全和废物管理方案）的共同立场报告，以使核能发电成为具有核能应用潜力的国家未来能源结构的一个组成部分。

就总体民用核能活动而言，法国和俄罗斯可以被认为是当下全球领先的国家。然而与此同时，中国在核电站建设方面正在取得重大突破，被认为是未来的潜在领先国家之一。

中法两国都认为核能可有效减少化石燃料的消耗，实现二氧化碳大幅度减排。在可预见的新核能技术角色和公众接受度方面，两国面临着类似的问题。在以打造"低碳"世界为目标的第 21 届联合国气候变化大会（COP21）的背景下，中法三院就向其他国家传达关于核能问题（其中主要涉及科学、技术、工业以及经济和社会等领域）的共同立场一事达成了一致。

这份共同立场报告就是中法联合研究小组的成果。

本报告涵盖核能的诸多领域，旨在为多个领域（核能在未来能源结构中的定位，核能的优缺点，研发前景，技术、安全和工程建设等）以及社会问题（教育、培训、风险认知、公众认识等）提供客观的综述，但也不可否认，本报告远非详尽无遗。比如，本报告未述及核电站寿期、运行延寿、退役等重要问题以及核能经济性方面的问题。此外，本报告也没有尝试将核能与所有发电方式及其各自的优缺点进行比较。

尽管专家组注意到当前某些核电站项目面临的难题，但是本研究并没有具体考虑这些问题，也没有尝试解决这些困难。人们认为这些难题终将是会被克服的，而本报告应将重点放在有助于开发和完成未来项目的方法和工具（特别是基于数字化的方法和工具）上。本研究无意介入当前正在进行的商务、管理、融资问题或商业谈判。

本研究旨在起草中法三院之间的共同声明，这些声明均载于本报告中（尽管中法两国的视角和国情并不完全相同）。

　　此外，本报告并非主要面向公众，而是在能源结构中关注核能发展的国家工作人员和职业人士，以及参加国际原子能机构（IAEA）年度大会的全球核能业界人士。在所探讨的某些问题上，专家组意识到需要做更多的工作来改善目前的现状。在社会问题上尤其如此，需向公众传达在寻找化石能源替代方案（寻找其他基本上不含碳、可运输且能确保稳定供应以满足需求的能源）时需要满足的诸多条件以及所遇到的巨大困难。为深入理解公众关注的问题，尚需做出进一步的努力，并创新方法。这方面的进展将取决于公众参与这些问题及其解决方案的讨论程度。然而，这些讨论超出了本报告的目的，因此还需要更多基于事实证据的经验和技术创新。

　　本报告于2017年7月定稿，作为此次联合研究的终场谢幕，中法三院在2017年9月20日召开的国际原子能机构（IAEA）年度大会的会外活动中向包括政策制定者在内的全球核能业内人士阐述他们的共同观点。此次演讲凸显了中法三院之间的合作，并期望通过提供更宽广、更长远的视角帮助认识当前西方核电项目所面临的一系列困难。

　　本报告由中法联合研究小组以中英法三语同时编写，现将中英双语版本正式出版。我们也将中法三院在IAEA大会中报告的演示文稿作为本书附件，读者可通过扫描封底二维码阅览。

Foreword

In the later part of the year 2016, the Chinese Academy of Engineering and the French Academies (National Academy of Technologies of France and French Academy of Sciences) set up a joint French-Chinese study group. Its general objective was to prepare a common position paper to distil policies and technology options, including safety and waste management, for making nuclear power generation a component of future energy mixes in countries with the appropriate potential for implementing this energy.

Regarding overall civil nuclear activities, France and Russia may, at this time, be considered the leading countries in the world. However, China is making impressive breakthroughs in setting up nuclear power plants and appears as one of the potential leaders in the future.

Both France and China are convinced that nuclear energy can effectively contribute to the reduction of fossil fuel consumption and consequently to a notable reduction in emissions of CO_2 in the atmosphere. They face similar issues with respect to the role of the foreseen new nuclear technologies and public acceptance. In the context of COP21, targeted at a "low-carbon" world, the three academies have agreed upon a common message addressed to other countries regarding nuclear issues, including mainly scientific, technological, industrial but also economical and societal aspects.

This common message is the outcome of the joint French-Chinese study group.

This report covers many aspects of nuclear energy and attempts to provide an objective overview of many aspects (position of nuclear energy in the future energy mix, benefits, strengths and weak points of nuclear energy, research and development perspectives, technology and safety, engineering etc.) and of societal issues (education, training, risk perception, public awareness etc.) but it is admittedly far from being exhaustive. For example, the important question of the life-time of nuclear power plants and the possible extension of their operation as well as questions pertaining to power plant decommissioning, as well as economic aspects of nuclear energy are not addressed. The report also does not attempt to deal comparatively with all the methods of energy production and with their respective merits and weaknesses.

Although the expert group is attentive to the problems encountered in some current nuclear power plant projects, these issues are not specifically considered in the present study and no attempt is made to examine these difficulties. It is considered that they will be overcome and that this joint report should focus on methods and tools that could improve the development and completion of future projects in particular those relying on digitalisation. There is also no intention whatsoever to interfere with on-going business, governance, financial issues or commercial discussions.

The study is aimed at developing common declarations between the partner academies, which are documented in the present report, whereas the visions and dynamics in China and France are not strictly the same.

Furthermore, this report is not primarily aimed at the general public. The target audience of the study is the "professional" public of countries considering the development of nuclear energy in their energy mix, and in particular the world nuclear community at the annual meeting of the International Atomic Energy Agency (IAEA). On certain issues, the expert group is conscious that more work needs to be done to improve the current state of the matter. This is true in particular for societal issues, where much more is required to inform the public about energy issues and about the extraordinary difficulties that will be encountered when trying to replace fossil fuels, finding other, essentially carbon free, sources of energy that can be mobilized and assuring a stable production responding to demand. To get a better understanding of public concerns will require further efforts, a point that is underlined but that will still need innovative approaches. Progress in this area will depend on the public's involvement in the discussions about these issues and their solutions. This debate goes, however, beyond the purpose of the present report and will need additional evidence-based experience and further technical innovations.

This report was finalized in July 2017. The three partners were bringing their joint work to a conclusion when exposing their common views to the world nuclear community, including policy makers, at a side event during the annual general assembly of the International Atomic Energy Agency (IAEA), on September 20, 2017. The presentation highlighted the present partnership between the Chinese and French academies and put into perspective some of the difficulties that are encountered in current Western projects by providing a broader, more long-term viewpoint.

目　　录

Contents

综述与建议

中国和法国都是拥有大型且持久核电站发电能力的国家。法国拥有长期运行核反应堆及所有闭式燃料循环设施的经验，而中国将是未来世界上新增核电装机容量最多的国家。两国目前都进入了优化能源结构的电气转型阶段，且均已开展针对下一代核电系统的研发项目。

基于对可再生能源的发展愿景，中法三院（作为独立机构）对核能所面临的一些问题和挑战进行了分析，并提出了以下想法和建议。

1. 实现能源体系的低碳减排面临诸多不应被低估的难题。通常认为，低碳减排要求增加电能在能源体系中的份额（例如：陆上交通、工业、城镇使用等）。核电是目前供应电能的最现实选项之一。近年来，核能发电在以安全、高效、清洁的方式供应电力的同时又解决了环境和气候变化问题，提供了一个极其现实的选择。核能能够可靠地供应可调度电力，对发电波动性强、不易调度以满足电力需求的间歇性可再生能源（如风能或太阳能等）形成了很好的补充。

2. 与能量密度更低的能源相比，大型核电站发电高度集中，因而所需厂区面积更小（占地面积更小），能为现今和未来涌现的特大城市提供所需的能源。核能消耗的原料铀是一种丰富的资源，由于不存在操控其市场供应的垄断利益集团，因此可以保证能源供应的安全。

（1）在可预见的未来，人类仍无法找到能够经济地储存大量电力的解决方案。因此，需要采取其他手段来补偿间歇性能源，以确保在出现更大的电力需求变化时，电力系统具备足够的反应和调度能力，而且不受时间限制。然而，如果利用化石燃料提供备用电力，那么低碳减排的目标就无法实现。

（2）间歇性能源在大型电网中的比重受技术和经济条件限制（电网稳定性、发生大规模停电时的恢复计划等）。

3. 核电仍是一种相对年轻的技术，目前还处于持续发展中。

（1）对重大事故进行了详细分析，总结的经验和汲取的教训使反应堆的设计运行得到了巨大改进。

（2）正在建设的最新一代反应堆（第三代）的设计可以保证即使发生堆芯熔化等严重事故，基本上也不会在核电站之外产生较大的放射性后果。

4. 其他技术发展路线已初见端倪，应鼓励其开发工作：

（1）第四代反应堆，包括钠冷快堆（SFR）。钠冷快堆可充分利用铀资源、实现乏燃料的多次循环和放射性废物中长半衰期放射性元素的嬗变。

（2）适用于小型电网的小型模块化反应堆（SMR），其模块化施工可降低大规模施工建设的复杂性。

5. 因此，维持现有规模的研发投入对降低成本和风险至关重要，其中包括：

（1）共享人力资源和研究基础设施。

（2）巩固和发展针对核电站的各个层面的设计、建设和运行的教育和培训规划。

6. 核能行业的健康发展有赖于权威的、独立的安全监管机构。

（1）它们可依托权威的、独立的技术支持机构（technical support organization，TSO）。

（2）这些监管机构、运营者和供应商之间必须开展技术对话，确保安全要求将科学技术知识的发展纳入考虑范围并结合对风险的客观评估。

（3）统一各国监管机构的安全要求是实现反应堆型号标准化的前提。这种标准化本身即可提高安全性。目前的现实情况是，各国安全监管机构所要求的安全水平不尽相同，有些国家实施成本－收益法，而其他国家则不接受这种方法——协调统一势在必行。中法三院建议采取"风险指引法"，使用该方法可平衡新的安全要求和效益。

7. 核工业的主要挑战在于：①废物管理；②成本控制、新项目融资；③向公众传递以下信息——核电站运行安全方面的进步已经达到相当高水平，理应得到社会认可。

（1）业界目前正在采用经过验证的公认的解决方案，管理低中放废物。目前已经拥有高放长寿命废物（HL-LLW）的整备技术（包括经过或不经过乏燃料后处理）。在利用谨慎选定的深地质层处置已整备高放长寿命废物方面，科技界还没有发现任何实质性障碍。处置场地的特征仍需进一步研究。

（2）核工业必须加强对核电项目成本和工期的控制。

对最近的一些项目所面临的困难，业主和供应商应进行认真的分析。中法三院建议公开这些分析的结论。

中法三院建议核工业应加快部署数字化技术。实践证明，数字化技术在其他行业的使用带来了诸多益处，这既有助于大大降低项目成本、避免工程延期，又能提升质量。其中有三大领域尤为重要：

①与设计过程耦合，应用并进一步开发仿真模型及平台的运用；

②仪控系统的数字化；

③采用具有唯一、统一数据库的数字化平台（产品生命周期管理，PLM），其中包含 3D 设计、施工、运行和生命周期管理等数据。

这些技术的引入需要安全监管机构的配合、法规的统一（特别是在网络安全领域）。

核电是一个生命周期很长的行业，需要大量的前期投资。因此，从国际投资银行获得充足资金对核电来说至关重要。

（3）如果得不到社会的接受和政府的支持，核工业的发展就会举步维艰。

应透彻地探讨安全问题，并向公众清晰地说明在核电运行方面过去与现在所做的改进。

能源方面的争议涉及复杂的技术和经济问题，只有依靠结构清晰的信息和可靠的数据才能正确处理这些问题。

对核工业的公正判断，迫切需要全面、客观的信息。必须倾听非政府组织的声音，但同时也必须听取运营者、设计人员、安全监管机构以及核工程和经济学方面的专家意见。政府应该征求专家的意见（包括国家科学院、工程院和技术院的专家意见）。

8. 核工业可通过国际和双边合作的形式惠及新兴国家。法国、中国等已经建设核基础设施的国家应帮助政局稳定的新兴国家以安全、高效的方式发展核电。同时，核工业界应在以下两个方面促进核电的发展：

（1）取证及监管流程的标准化和稳定化；

（2）实现有政府担保的长期购电协议（PPA）的普遍化。

第1章 简介：核能在未来能源结构中的必要性以及需要克服的挑战

作为社会经济发展的支柱，能源是人类生活、健康、福祉不可或缺的必需品。满足全球的能源需求、减少全球温室气体的排放量，是全球未来面临的根本性挑战。

自20世纪50年代以来，和平利用核能已经成为在不依赖于煤、石油和天然气等化石燃料的情况下提供能量的一种可行方案。近年来，核能也已被视为一种能保证能源供应、支撑全球经济发展同时又能减少温室气体排放的可持续和可靠的低碳能源。虽然核能已经成为全球电力供应的三大支柱之一，但其发展正处于关键时期。日本的福岛核事故引发了巨大的恐慌，让公众对核安全心生疑虑。出于该原因以及一些其他的原因，有些国家已放弃使用核电、关停了现有的核电站。吸取福岛核事故的经验教训之后，对现有的核反应堆制订了重要的安全改进计划，未来的设施也将得到进一步的改进，因此如今依然有多个国家在维持核电的发展。

从中长期来看，要显著减少导致气候变化的温室气体排放、减少氮氧化物/未燃烃/颗粒物等污染物的排放、改善大气环境、实现人类的可持续发展，全球能源体系应更加环保，且尽可能少地利用化石燃料。在当前大部分电力依靠化石燃料——更确切地说依靠煤炭生产的大背景下，核电是以安全、高效、清洁的方式供应能源，同时又是解决环境和气候变化问题的最现实选择之一。

由于核能是一种稳定的能源，因此它能可靠地供应基本负荷电力，并对波动性较大、不易于调配以满足需求的间歇性可再生能源（如风能或太阳能等）形成了很好的补充。在这方面，人们普遍认为，且有最近经验表明，在大多数国家，间歇性可再生能源在电力结构中的总体比重不能超过30%～40%，否则会引起不可接受的电力成本，导致温室气体排放量的增加，以及产生电力供应安全风险。可再生能源应用的瓶颈主要在于无法大规模储存电能，目前仍没有任何迹象表明该技术难题在未来会被攻破。与集中度低的能源资源相比，核电采用的是大型的集中式电站，其占地面积更小，因此它能提供现有及未来特大城市所需要的能源。

然而，发展核电在安全性、放射性废物管理、先进核能系统的开发与部署、经济性、公众的接受程度等方面依然面临着诸多挑战和问题。作为具有强大核电站建设能力的主要国家，中国和法国都非常重视全球核能的和平利用，并且有责任、有意愿帮助新兴国家发展核电，共同应对他们所面临的挑战。

继致力于大幅减少全球温室气体排放的第 21 届和第 22 届联合国气候变化大会之后，中法三院相信，其围绕核电有关复杂问题提出的倡议能为其他国家的科研机构、决策者乃至社会各界提供有价值的信息。

本研究中的立场反映的是中法三院作为独立机构的立场，不应被解释为核电从业者的立场，或被解释为法国政府、中国政府的立场。

在本报告中，中法三院旨在勾勒出核能的历史与前景，解决需要考虑的关键问题，以让核能变得更安全、更经济，惠及发达国家和新兴国家。

虽然这篇报告提及了诸多问题，但其并非是包罗万象的。它代表了过去 6 个月（2016 年 11 月至 2017 年 4 月的报告编制讨论期间）我们的思考和讨论成果。

本报告共包含 17 章和一个综述。第 2 章和第 3 章简要介绍了核能开发的历史、问题与挑战，并讨论了与第三代核电站部署有关的问题。第 4 章到第 6 章从科学的角度论述了未来反应堆设计的前景与挑战，具体讨论了第四代反应堆的现状以及小型模块化反应堆理念。第 7 章和第 8 章讨论了技术方面的问题，包括安全问题、对技术支持机构的需要、数字化以及新型设计工具的进步与挑战。第 9 章强调了核研究设施和基础设施的重要性。

第 10 章论述了人员的教育与培训问题。一个重要的方面是吸引年轻的高校毕业生进入核工业；另一方面是培训企业员工，赋予他们适当的科学教育背景，让他们了解安全管理文化。在这方面则强调了模拟机的作用。

第 11 章和第 12 章讨论了核电项目管理等工程问题，以及在满足安全需求的前提下控制成本与工程复杂度的问题。第 13 章包含了关于为新兴国家核项目提供国际支持的内容。

第 14 章到第 17 章讨论的是社会问题。我们对过去 50 年来全球核活动对人类健康的影响进行了评估（第 14 章）。由于安全是核电站运行过程中的核心问题，因此有必要审视对风险的认知及其与实际危险的关系（第 15 章）。第 16 章强调了提升公众认识的必要性，并思考了为达到此目的而所需要进行的治理。第 17 章论述了提高公众理解的组织形式、方法和不同利益相关方的责任，并且还讨论了提高公众理解需要采取的行动及其必要性，以降低因加大监管力度而导致核设施开发和运行复杂化的风险。

第 2 章 核能开发的历史、现状、问题与挑战

建议

现阶段，为使核能在未来各国能源结构中作为低碳能源发挥显著作用，提出以下主要建议：

● 为了让公众恢复对有能力驾驭核电复杂性的核大国的信心，核行业应当努力在预算范围内按时完成核电项目。为了达到这个目标，有必要认清在一些西方国家中三代＋核电站工程建设表现欠佳的原因，与此同时，采用同类技术的工程建设在中国却按计划实施。这些分析的参与者需包括项目的投资方、业主、监管机构、设计方、工程团队、承包商和分包商等。

● 尽可能加快小型模块化反应堆（SMR）的商用部署速度，展示更加灵活的核反应堆应用前景：即使是电网容量有限的地区或新兴国家也可以使用。

● 有核国家的政府和金融市场参与者应共同重新审视核电项目的融资机制。

民用核能是人类历史上面临的最大技术挑战之一，因为它涉及方方面面，各方面都需要非常专业的知识以及容错模式等手段，包括安全、设计、施工、试运行、项目管理、法律和监管问题、废物管理和停运、融资、安保与核不扩散问题、有害辐射防护、环境影响、公众的接受程度等。

自 20 世纪四五十年代核能问世以来，核工程变得愈发复杂，而实现"设计项目"所需的科技和工业能力是唯一需要克服的挑战。

自那时起，得益于在反应堆设计和研究堆、实验室、核燃料制造和后处理设施等其他核设施的稳步发展，民用核能得到了不断发展。三代反应堆技术相继问世，有些

方面是重复的，设计难度不大，每一代的安全性至少提升 10 倍。

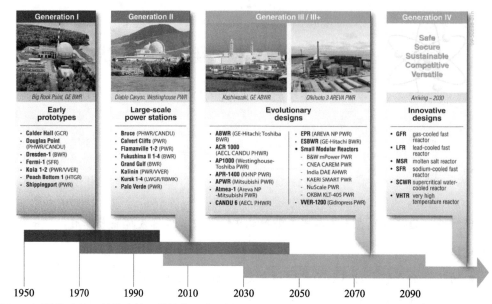

图 1　历代核电站。转载自第四代核能系统国际论坛（www.gen-Ⅳ.org）——图中未显示中国的第三代/三代＋核反应堆：HPR1000（CNNC PWR），CAP1400（SNPTC PWR）和 ACP100（CNNC SMR PWR）

鉴于三大核事故（三里岛、切尔诺贝利和福岛）和"9·11"恐怖袭击事件带给我们的经验教训，三代＋核电技术必须满足核辐射安全的高标准，但这也导致成本增加，降低了核能的竞争力。

第四代核电技术目前处于筹备阶段，很可能会在 2030 年之后推广应用，它旨在解决核裂变材料的经济性问题以及核燃料闭式循环问题。

在西方国家，由于切尔诺贝利核事故的发生，核能领域一度遭受冷遇，从 2003 到 2016 年，核能的发展一直停滞不前，几乎没有启动新的核电项目，核电站技术人员也少有新鲜血液加入。这对显著增强了安全性能、具备更加安全的设计和更加复杂的审批流程的第三代和三代＋核电项目的发展带来了严重影响。

幸运的是，下列两个因素在解决这些难题的方面发挥了很大的作用：

● 供应商在模块化建设和可靠供应链的重启方面进行投资；

● 韩国和中国等新客户/供应商的加入，他们拥有大量"全新视角"、富有激情且训练有素的核电人才。

主流反应堆方面，压水反应堆逐渐在商业上战胜竞争对手：首先淘汰了重水反应堆，接着在福岛核事故之后，淘汰了沸水反应堆——尽管这两种技术也有各自的独特优势。

与主流反应堆类似的是，核燃料循环设施和（更普遍的是）前后端行业的设计和施工至今没有遇到较大的困难，除铀浓缩设施采用了离心技术外，尚未出现明显的更新换代。必须提及的一点是主要锕系元素的分离/嬗变工业流程和技术已经过研发论证，但在产业规模方面属于第四代专业反应堆或加速器驱动次临界系统。

现有问题与挑战大多与核电项目的特殊环境有关。可归纳为三大问题，每一个问题都会产生严重的直接后果，阻碍核电事业的发展，它们相互之间的影响很大，如果不处理其他两个问题，那么任何一个问题都无法得到解决！这些问题是：

● 世界各地的公众对大型项目（尤其是核电项目）的信心在下降，因此需要制定更严格的安全要求；

● 很多国家的审批和控制流程日趋复杂；

● 由于成本上涨和愈发谨慎的银行规则，新核电项目的融资难度（政府资金和民间融资）越来越大。

就这些问题而言，有些决策者（尤其是欧洲的决策者）认为，只有国有供应商或电力公司能够开展核电项目。人们不禁思考这是否是向脱碳能源系统转型的最佳解决方案，且在安全和成本问题上，私有公司之间市场竞争的模式也无法在安全和经济方面产生更优化的结果。

对高放射性长寿命的核废物的长期管理依然是一个悬而未决的关键技术问题，该问题受到了众多国家的广泛关注。根据法国在两部法律（分别于 1991 年和 2006 年颁布）框架下开展的研究，我们相信有助于减少核废物的解决方案是存在的，它通过规范快中子反应堆的建设和运营，将次锕系元素转化为寿命约 300 年的裂变产物——远低于次锕系元素。因此，我们未来可能对进行过后处理的核废物持有更加乐观的态度。但目前最切合实际的解决方案是将现有的废物和乏燃料存储在深地质处置库中。目前这类在建处置库的案例较少，因此出现的相关技术问题也很少，现有问题也正在得到解决，但与之相关的公众接受度的问题依然需要引起重视。

未来，提升公众对核电的信心的关键在于向越来越多的国家验证第一代和第二代核电站以及其他核设施的退役可行性。技术和安全水平的提升必不可少，但对公众的信心而言，更重要的是在预算范围内按时完成核电项目建设。

结论

最重要的是向决策者和公众解释：核能能够有效地控制全球变暖，经证实，核能是实现充足、安全可靠的基本负荷供电的最有效途径之一。

其次，将核活动与其他存在具体风险的人类活动所达到的安全水平进行比较也同样重要，并解释核领域是如何显著降低风险的。第三代核电站以及福岛核事故后的增强型第二代核电站的安全设计已排除了在发生严重事故情况下采取应对措施的必要性。换言之，这些反应堆在发生严重事故时不会造成严重的放射性后果，因为所有可能的排放都将被限制在反应堆厂房中。

值得一提的是，大多数西方国家的第二代反应堆的使用寿命将在 21 世纪中叶之前结束，届时将需要进行更新。

第 3 章　近中期部署第三代核电站的可行性与挑战

建议

通过优化设计，第三代或三代＋核电站的经济竞争力应得到进一步提升（包括数字化、模块化和标准化施工等；同时还包括创新融资解决方案以及更多的支持性法规）。法国二代＋反应堆的标准化策略有助于提高安全性，同时降低建设成本。中国在设计与工程的数字化和核项目管理方面已取得长足进步。两国分享各自的经验将有益于未来世界核工业的发展。

世界上目前并网的核反应堆主要是第二代和改进型第二代（二代＋）反应堆。它们的基本设计是在三里岛 2 号反应堆事故发生前完成的，然而大多都增加了安全保障手段，以吸取三里岛 2 号反应堆事故（反应堆安全阀、人机界面等）、切尔诺贝利事故（安全壳内氢复合器、安全壳过滤通风系统）的教训。在福岛核事故发生前后，第二代反应堆针对全厂断电工况增加了能动和非能动安全系统，针对最终热阱丧失工况的排热也进行了改进。值得注意的是，日本的第二代核电站基本没有实施这些改进措施。另外，这些第二代反应堆经改进后可以达到改进型第二代反应堆的标准，可以认为它们在安全水平方面与后来反应堆接近。

20 世纪 90 年代初，从德国和法国开始，安全监管机构要求新建反应堆应对内部事件时应满足下列安全目标：

● 必须实际消除可能导致早期或大量放射性泄漏的堆芯熔化事故。

● 对无法完全排除堆芯熔化可能性的严重事故，必须设计预案，保证只需对公众在一定地域/时期内采取有限保护措施（无需永久迁居、核电站周边地区无需紧急撤离、只需为有限的人员提供庇护所、无需长期限制食品消费），且必须通过指定预案，保证有足够的时间来实施这些措施。

● 远在福岛核事故发生之前，国际核安全咨询组（INSAG）、国际原子能机构总干事的高级咨询小组，在其 INSAG-12 文件（1999 年）中提出了非常类似的安全目标①。虽然这些要求没有被纳入国际原子能机构的《基本安全原则》②，但被收录到了国际原子能机构的安全标准《核电站安全：设计》③ 中，其与西欧核能监管机构协会的安全目标④一致。

在外部事件方面，如今的监管机构倾向于要求将大飞机蓄意撞击考虑进去④，并要求证明这种情况下反应堆能够安全停堆。同时，还必须考虑超设计的外部危害（地震、洪水等），以证明陡边效应不会严重影响核安全。

除改进型第二代反应堆采用的安全改进措施外，三代＋反应堆还增设了反应堆堆腔熔融物收集装置（堆芯捕集器），或者设置防止反应堆压力容器熔穿的系统、冷却熔化堆芯的安全壳内（或堆内）换料水箱以及一系列备用电源。要缓解类似"9·11"恐怖袭击的坠机所造成的事故后果，要求设置双层安全壳，并配备相应的冷却系统。

所有这些安全手段结合在一起，使反应堆高压熔融事故频率与二代＋相比降低90％以上，并保证事故发生时几乎所有的安全功能均基本可控。通过实施长期操作员干预战略减少了人因失误，大大减轻严重事故的放射性后果。"无需永久迁居、核电站周边地区无需紧急撤离、有限的人员庇护、无需长期的食品消费限制"的目标业已实现。

从运行的角度来看，到达堆容器的中子通量减少，为延寿创造了实质性条件。最后，通过明智地选择了易于清理和维护的材料、设备，工人的总照射剂量也降低了。与第二代相比，第三代反应堆每太瓦时（TWh）的核废物将会大大减少。

美国和欧洲的公用事业机构将监管要求转化成了实操性更强的规格书（美国的URD 和欧洲的 EUR）。基于这些要求，业界已经设计出了一些堆型，如 AP600/1000、

---

① 国际核安全咨询组. 核电站的基本安全原则-INSAG-3 Rev. 1 1999 年修订版.
② 国际原子能机构. 基本安全原则-安全要素-SF-1-2006.
③ 国际原子能机构. 核电站安全：设计-特殊的安全要求-SSR-2/1（Rev.1）-2016.
④ 西欧核能监管机构协会. 新 NPP 设计的安全性报告（2013 年）.

EPR（欧洲压水反应堆）、VVER1000/1200、华龙一号（HPR1000）、CAP1400、APR1400、APWR1500、ABWR、ESBWR 等。

满足这些要求并达到第三代标准的反应堆已经满足投入工业应用的要求。1996年以来有两个先进沸水堆机组已投入商业运行。有八个 AP1000 机组正在中国和美国建设或试运行，其中示范项目位于中国三门。四个 EPR 机组分别位于芬兰、法国和中国，其首堆位于中国台山。中国于 2015 年开始建设四个华龙一号机组。三代＋机组将是近中期大规模部署的核电主力机型。

由于安全性的提升以及首堆工程实施的复杂性，到目前为止，除华龙一号外，几乎所有第三代示范项目都遭遇了工期延长，或因成本上升导致超出预算和融资压力，项目不得不推迟。

由于在安全方面投入巨大，第三代/三代＋核电站的经济性竞争力必须进一步提升，包括通过优化反应堆设计、利用现代信息技术、大宗采购、模块化和标准化施工、全球供应链的合作、创新融资解决方案、支持性的监管和监督环境，以求最大程度缩短工期，提高热效率和反应堆利用率，继而提高核电的安全性和竞争力。

法国二代＋反应堆的标准化战略，在降低建设成本的同时，为安全性提升做出了重要贡献。这一战略同样可以用于降低第三代核电站的成本，使核电继续成为最具竞争力的基荷来源。关于用第三代反应堆逐渐替代法国部分二代＋反应堆的计划目前正在研究论证当中。

中国制订了发展核电的宏伟计划，得益于过去 30 年核电站的持续建设，中国已建立起强大的核电基础设施，包括研发、设计和工程、制造、建设和安装、项目管理、调试和运行能力等。在此基础之上，中国计划与相关方开展战略性合作。

结论

大部分并网的核反应堆都属于第二代反应堆，其中很大比例已改造升级为第二代＋反应堆，这些反应堆能够满足更严格的安全要求。三代＋反应堆包含在堆芯熔毁时收集熔融物的堆芯捕集器以及其他安全设施，可以使高压堆芯熔毁的频率降低

90％以上。操作员干预策略在事故下可留出足够时间采取行动，提高人为失误容错度，从而使核电站附近大范围居民无需撤离，也无需担心食物受到污染。三代＋反应堆在压力容器处的中子通量和控制工人照射剂量方面也已取得类似进步。目前，三代＋反应堆已能投入工业应用。然而，安全性的提升也使三代＋反应堆付出了代价，如工期延长、预算超支和竞争力下降等。

第 4 章　新型和创新型未来反应堆设计 (第四代反应堆) 的前景与挑战

建议

为及时准备好第四代钠冷快堆 (SFR) 和超高温气冷堆 (VHTR) 的商业化运营，中法三院建议在未来几十年里建设和运行以上第四代堆型的示范堆，以及燃料循环的相关设施。启动和运行 SFR 要求对乏燃料进行后处理，循环利用钚和铀。因此，应尤其重视 SFR 的燃料循环。鉴于当前的工业发展水平，SFR 是一种有前景的技术，有望利用自裂变核能时代起积累至今的裂变/增殖核材料提供电能。VHTR 则具有提供具有多种工业用途的高温热能的潜力。根据目前的各国能源战略，第四代 SFR 和 VHTR 的部署将晚于 2030 年。

核电的一个核心问题是如何最大限度地从核燃料中提取能量，同时最大限度地保证安全。要做到这一点，需要使用铀和钍的可裂变和可增殖同位素。可增殖同位素 (如铀 238) 通过与热中子或快中子作用诱发核反应，产生易裂变同位素 (如钚 239)。

4.1　现状

今天，全球的核电 (除少数机组外) 都是由装载天然铀经富集铀 235 后制成的铀氧化物燃料 (UO_x 燃料) 的热堆生产的。该燃料释放的能量来自于铀 235 (大约 70%) 和钚 239 的裂变，其中钚同位素是铀 238 在燃料中就地转化而成的。

有一些热堆部分装载了混合铀－钚氧化物燃料 (MOX 燃料)。其中用到的钚是

通过后处理从 UO_x 乏燃料中提取出来的，而铀是天然铀浓缩后剩下的贫铀。MOX 燃料中的能量主要来自于钚裂变（90％）。回收利用 UO_x 乏燃料中的钚可以减少天然铀的消耗量和分离功的运行，但即便在合适的热堆中，也只有第一次能相对容易地回收利用钚。

热堆没有充分利用天然铀的潜能，因为几乎所有的铀 238 同位素都残留在乏燃料中。另外，出于安全方面的原因，热堆中燃料的燃耗一般限于 50 GW·d/t（吉瓦·天/吨），冷却剂的温度仅为 300 ℃左右。

使铀 238 裂变或将其高效地嬗变成钚 239 的唯一办法就是使用快中子。快中子反应堆的设计就利用了这种中子特性。能量来自于铀和钚所有同位素的裂变反应。快堆中 MOX 燃料的燃耗可高达 140 GW·d/t 或以上，冷却剂的温度可达 500 ℃。

目前，世界上只有俄罗斯有几个快中子反应堆机组正在并网发电。

4.2　主要预期变化

2002 年，第四代核能系统国际论坛（GIF）发起了有关未来核能系统的联合研究。中、法、韩、日、俄、美、欧盟之间由此展开了积极合作。GIF 提出了六大领域的技术目标和相关评估指标：可持续性、经济性、安全与可靠性、废物最小化、防扩散和实体保护。六类最有前景的核系统被选中，其中两类为气体（氦）冷却反应堆，另两类是液态金属（钠、铅合金）冷却堆，还有一类超临界水冷堆，最后一类是熔盐冷却堆。

在这些被选中的反应堆系统中，几乎所有的 GIF 合作国都认为，使用 MOX 燃料的先进钠冷快堆（SFR）在 21 世纪投入商用的可能性最大。中、日、韩紧随美、德、南非的步伐，正在积极开发超高温气冷堆（VHTR）系统。虽然法国对 VHTR 的开发做出的贡献有限，但已经启动了气冷快堆（GFR）的研究。

钠冷快堆主要用于发电，自 1951 年以来已经积累了 400 多堆年的运行经验。超高温气冷堆则主要用于热电联供和氢气生产。氢是通过热化学、电化学或混合工艺从水中制取的。超高温气冷堆因出口温度高，对化工、石油和钢铁工业也很有吸引力。运行超高温气冷堆的经验从 1963 年就开始积累了。

钠冷快堆和超高温气冷堆都具有潜在的固有安全性，其设计旨在任何运行或事故工况下排除放射性裂变产物释放到环境中。

基于自身运行钠冷快堆的经验，结合 MOX 乏燃料后处理和钚循环利用的工业基础，法国正在积极开发一座 600 MWe 的第四代钠冷快堆。中国计划根据中国实验快堆方面的经验，开发 600 MWe 示范钠冷快堆。日本也有运行钠冷快堆的经验。俄罗斯已实现了快中子反应堆的并网发电，并已制订长期发展计划。印度的民用核计划虽然迄今为止基于钍，但也包括建设使用 MOX 燃料的钠冷快堆。

中国目前还在实施一项具体的超高温气冷堆长期研发计划。

4.3　第四代钠冷快堆和超高温气冷堆的前景

目前，从很多方面来看，第四代钠冷快堆前景广阔。其中的一个重要方面是，几乎所有铀和钍的同位素和重原子核在一定的快中子通量下均可发生裂变。这意味着贫铀以及乏燃料中的铀和钚可在快堆中多次循环利用。只要乏燃料从反应堆中卸出后很快得到后处理，快堆就可产生自己的 MOX 燃料。钠冷快堆 MOX 燃料中，15％的铀238（来自富集后余下的贫铀（DU）或后处理得到的铀）被转化为钚。用快中子反应堆从天然铀中提取能量，仅会受到工业化循环乏燃料能力的制约。因此，有了快中子技术，可裂变物质中的资源几乎是无限的。

更高的燃耗使得钠冷快堆燃料在堆中停留的时间达到热堆中的两倍，同时也降低了乏燃料中次锕系核素的含量。因此，钠冷快堆运行起来比热堆更经济，并为废物管理提供了更光明的前景。与热堆相比，钠冷快堆产生的高放废物含有的长寿命放射性核素更少。这一点将在专门讨论核废物管理的章节中予以论述。

按生成钚的量/速率，钠冷快堆可以有多种配置，可以设计为生成钚与其使用的一样多（也可以更多或更少）。在生成等量钚的配置中，钠冷快堆中钚的量保持稳定，在燃烧器的配置中，可大大减少钚的库存量，从而降低核扩散风险，以便在未来某个时期决策，安全地规划核能发电的终结。钠冷快堆还可设计用来嬗变镅等超钚元素。

在启动第四代钠冷快堆系列项目前，需要解决在此类反应堆的布置和掌握相应的燃料闭式循环方面的各种科学和技术问题。

2014 年更新的第四代技术路线图显示，超高温气冷堆可在 700～950 ℃（未来还可能超过 1000 ℃）的堆芯出口温度范围内供应核热和电力。堆芯出口温度指氦的温度，即一次冷却剂的温度。至于水蒸气的温度，则取决于特定的设计。超高温气冷堆的研发关注的是燃料和燃料循环，材料，制氢，仿真、验证和基准对比的计算方法，部件和高性能涡轮机械，系统集成与评估等。

第四代反应堆的安全水平至少与第三代热堆相当。这需要对传统快堆的设备进行很多革新。根据第四代核能系统国际论坛的技术路线图，在法国和中国，这种反应堆的技术性示范堆的设计目前正在进行中。钠冷快堆应被视为核能未来发展蓝图中的一部分。该技术有潜力提供几乎无穷无尽的可调度电力资源，因为它能嬗变/分裂所有铀同位素（而不只是自然铀中 0.7％的铀 235）。钠冷快堆也能帮助控制钚库存和嬗变长寿命锕系元素，这对长寿命核废物的管理有利。法国曾是该技术的先驱者，如今又通过结合过去反应堆的运营反馈和重要创新，致力于开发商用反应堆一半功率的技术示范堆（Astrid-600 MWe）。中国已从 2011 年开始成功地运行钠冷快堆原型堆（20 MWe），在 2017 年底开始建设 CFR600（同样也是商用反应堆一半功率的示范堆）。中国首座商用 SFR 最早能在 2035 年开始建设。法国预计将在 21 世纪下半叶部署商用 SFR。

结论

总而言之，第四代钠冷快堆和超高温气冷堆是在支撑未来核电能源发展、提供电力或热（高温流体/气体）电联产方面最有前景的反应堆。在当前的发展阶段，核能生产主要基于热堆，发电量的增长也增加了全球的钚 239 库存，速率达 75 吨/年，其他核材料同样如此。钚库存的大量增加引发了核不扩散问题。当乏燃料被认为是废物时，出现乏燃料的处置问题；当被认为是资源时，就存在人造核裂变材料在热堆中的利用问题等。很多国家在今后数年仍会产生大量的钚，为在未来几十年内启动安全性提升的第四代钠冷快堆开创了前景：钚库存的利用将供应可用数百年的能量。这同样意味着需要大大改进目前的反应堆系统，建立一种新型核燃料循环体系，并将重点放在核能长期利用的研发工作上。

附录 A

A.1　法国第四代反应堆行动计划

法国原子能和替代能源委员会从 2010 年开始负责管理先进钠冷技术工业示范反应堆项目 Astrid。该项目目前包含两个部分：(i) 反应堆设计；(ii) 用于生产 Astrid 专用的首炉 MOX 燃料以及回收利用钚和测试镅嬗变的装置设计。Astrid 还必须证明钚的燃烧是可实现的。该项目聚集了核能领域内外的很多行业合作伙伴，吸收了 40 年来运行 Phenix 和 Superphenix 快堆的经验。

Astrid 将被设计为装载 MOX 燃料的池式钠冷快堆，600 MWe 的功率将达到工业示范堆的标准。其一回路将具有四个环路，二回路中每个钠环路设有一个蒸气发生器或氮气发生器（取决于不同的设计版本）。蒸气和氮气参数约为 15 MPa，500 ℃。反应性控制由两套停堆系统、一套独立补充停堆系统实现。热池和堆容器与多套余热导出系统相连。

Astrid 在安全方面的创新主要集中在失钠事故下对堆芯反应性的控制，以及在所有情况下的冷却和对放射性的限制。法国原子能源委员会已取得空穴系数为负但趋近于零的非均质新式堆芯的专利，这是快堆技术的一大进步，需要专用的 MOX 燃料子组件才可实现。目前在役的钠冷快堆中没有一个具有此类固有安全性的堆芯。蒸气发生器的设计改进使得钠水反应的主要化学风险得以控制。另一种方案中，目前正在对一种可以消除钠水接触的钠氮热交换器进行测试。这种钠氮热交换器与氮气汽轮机一起，将成为第二大创新点。

法国原子能委员会还取得了新的钠泄漏检测方法的专利。如堆芯完全熔化，堆芯熔融物会被回收。余热排出由几个非能动系统保证。最后，如有事故发生，双层安全壳和气密系统将避免任何放射性物质释放到环境中。

可运行性方面的创新涉及新的在役检查系统和反应堆部件维修。为节省操作时间，针对燃料子组件的特殊换料系统正在设计当中。

为 Astrid 制备新 MOX 燃料的工艺是工业可行的。要解决的主要难题是对钚含量较高的钠冷快堆 MOX 乏燃料进行工业化后处理,将周转期缩短为几年。法国已有数千吨 UO_x 乏燃料的后处理经验。

虽然钠冷快堆的设计在过去二十年内取得了很大的进步,Astrid 仍需进一步研发,改进反应堆部件、燃料包壳材料等。例如:可将燃耗提高到 200 GW·d/t 的包壳钢材就具有特殊的研发价值。

启动 Astrid 建设的决策将不会在 2024 年以前做出。法国未来的核能战略包含通过储存未经后处理的热堆 MOX 乏燃料形成钚的战略储备,该储备中也包括贫铀等其他核材料。这让自给自足的钠冷快堆系列在 2050 年后(最乐观的情况)有条件逐渐启动,很有可能也能嬗变自身将产生的镅。

同样,用铀 235 或钚作易裂变同位素、启动可增殖单一同位素钍 232 的热中子反应堆也是有可能的。易裂变铀 233 是通过钍 232 的核反应过程在燃料中生成的。铀 233 有吸引人的易裂变属性,并且它能或有可能通过钍 232 的辐照,在热堆中生成。利用钍的原型堆或多或少取得了一些成功。印度对"钍循环"进行了实验。法国正在研究建设的现代版快中子熔盐反应堆(FRMSR)具有高运行温度(600~700 ℃),其中熔盐(钍、锂和铍的氟化物)既是燃料又是冷却剂。该系统生成重锕系元素很少,但必须通过批量或在线的高温化学工艺定期从燃料中提取裂变产物。同样,还需提取的物质有铀 233,以保证快中子通量稳定。该反应堆可在工作中嬗变锕系元素。

钍基熔盐堆技术仍有很多问题有待解决。虽然液体燃料有其吸引力,但又提出了新的问题:余热冷却、腐蚀、放射性包容、高能伽马射线防护和新型废物管理。熔盐反应堆的优缺点相当,在法国预计不会大规模用于发电,除非大量的可转换核素变得必要。

法国对超高温气冷堆开发的贡献一直局限于材料和制氢。作为第四代核能系统国际论坛研究的一部分,2000 年到 2008 年间,法国原子能委员会为设计气冷堆模型进行了大量工作。Allegro 计划设计一个功率为 75 MWth 的动力堆,用氦气冷却,压力 75 bar,温度 850 ℃,首炉堆芯采用 MOX 燃料(含 30% 钚,不锈钢包壳),后续长期将采用一种 UPuC 混合碳化物堆芯,钚的含量不变,但改用碳化硅包壳。2005 年,法国原子能委员会将 Astrid 的研发放在了优先位置,但 Allegro 项目依然是欧洲规划的一部分,只是功率降到了 40 MWth,这是由于目前尚无通过鉴定的碳化物燃料,而材料也需要进一步研发。

A. 2　中国第四代反应堆行动计划

1. 钠冷快堆

作为中国核能发展战略"热堆-快堆-聚变堆"的第二步，开发快堆的主要目的是满足能源需求、缓解天然铀资源的潜在短缺。另一个目的是嬗变长寿命核素。

中国将在 2025 年之前建成钠冷快堆示范堆 CFR600。CFR600 的目的是示范燃料闭路循环，为大型钠冷快堆制定标准和规范。

CFR600 的技术选项基于以下目标：

- 满足福岛核事故之后修订的国家核电设计安全规定。

- 运用固有安全特性和纵深防御原则。力争达到第四代核能系统的可靠性与安全要求。

- 采用成熟的、技术上经过论证的部件。所有创新设计都应经过不同方法的充分论证。

- 与中国实验快堆和其他钠冷快堆项目比较，努力提高其经济性指标。

CFR600 将设计为采用 MOX 燃料的池式快堆。其热功率为 1500 MWth，电功率为 600 MWe。一回路中有两个环路，二回路的每个环路有 8 个模块化蒸气发生器，三回路是安装了一个汽轮机的典型水-蒸气系统。蒸气的参数为 14 MPa、480 ℃。反应性控制由两套停堆系统、一套独立补充停堆系统实现。一套非能动余热导出系统与热池相连。主次安全壳专为该反应堆设计。CFR600 的初步设计于 2016 年完成，2017 年底浇灌第一罐混凝土（FCD）。

2. 超高温气冷堆

中国于 20 世纪 70 年代中期开始研发高温气冷堆，HTR-10 高温气冷堆实验堆于 20 世纪 90 年代建成。现在正在开发"高温气冷堆球床模块"HTR-PM，这是一个业内技术领先的示范项目。2008 年 2 月，作为国家科技重大专项的 200 MW HTR-PM 示范电站获批。项目的路线图报告显示，在中国开发 HTR-PM 的前景在于其作为压水堆的补充，可以提供高效的核工艺热能的潜力。HTR-PM 的开发还将对先进核技术领域的创新做出更多贡献。HTR-PM 示范电站由两个球床反应堆模块组成，外加一个 210 MWe 的汽轮机组。反应堆堆芯入口/出口的氦气温度分别为 250 ℃/750 ℃，蒸气发生器出口的蒸气参数为 13.25 MPa/567 ℃。HTR-PM 示范电站于 2012 年 12 月 9 日在山东省荣成动工，反应堆厂房的建造于 2015 年完工，两台反应堆压力容器于 2016 年安装到位。2005 年，一条原型燃料元件生产线在清华大学核研院（INET）建成，每年可生产 10 万个燃料元件。此后，一个具备年产 30 万个燃料元件产能的燃料元件厂在中国北方的包头建成。HTR-PM 示范电站用的燃料元件于 2016 年开始生产。预计继 HTR-PM 示范电站后，将通过批量建设的方法对 HTR-PM 进行商业部署。带有更多（例如：六个）模块的机组正在设计当中。目前还在研究开发带有多个与一台汽轮机相连接的标准反应堆模块的核电机组。

第5章 创新型反应堆概念和技术的承诺与挑战：小型模块化反应堆及先进技术

建议

轻水反应堆是一项具有巨大改进潜力的朝阳技术。

● 应鼓励发展小型模块化反应堆，并为其提供财政支持，因为它们是解决低碳经济需求的一种灵活手段。应进一步加强 IAEA 框架内的国际合作，以制定出可从一个国家移植到另一个国家的新监管机制。

● 应通过研发获取适用于所有轻水反应堆的期望成果。核相关技术的研发投资应维持在较高水平上。应系统性地促进来自其他行业的技术转让。借助 IAEA 以及美国核管会所采取的一系列行动，监管部门在审核新性能和新技术时，应充分考虑在其他行业使用该技术时获得的经验教训。

在能源转型的背景下，市场的需求又让大家开始重新关注小型模块化反应堆。在低碳经济时代，它们可以找到自己的一席之地。小型模块化反应堆可成为清洁稳定的分布式能源。

SMR 的发展正好能够证明，轻水反应堆技术并未冻结。研发目前仍在继续，并且结果将用于简化核反应堆设计和运行（不论其规模），同时提升它们的安全性。

5.1 小型模块化反应堆

5.1.1 研发背景

自 20 世纪 70 年代以来，主要出于经济方面的原因，基于传统环路式反应堆核电站产业发展伴随着反应堆发电能力的持续提升（最新型号的反应堆发电能力已超过 1500 MWe）。

此外，还推进了创新的中小型反应堆（50～300 MWe）的研发，以支持多种应用：偏远位置、未并入邻近电网的中小型电网、旧燃煤机组更换、将废热用于供热网络的热电联产（集中供热和工艺加热供应）、海水淡化、海岛开发、将核电项目逐步引入新的国家。

显而易见的是，尽管市场前景广阔，也开展了一些设计研究，但这些小型模块化反应堆还未能展示出大规模应用的实用性，这主要是由于开发的进展缓慢、经济方面的原因（安装、分散和培训成本等），以及厂址准备和部署的延迟。

5.1.2 重拾兴趣

由于拥有先进的领导地位以及在能源领域、核技术方面取得的重大进展，在美国能源部的倡议下，近年来，大家对小型模块化反应堆的兴趣越来越浓厚。小型模块化反应堆是游戏改变者，能够以高安全水平提供不同的核电联产解决方案。

在能源方面，开发小型模块化反应堆的主要原因可归结为四种：①减少化石燃料使用的需求；②分布式能源（可再生能源、智能电网、储能）；③要求操作人员机动、灵活；④长期投资项目的融资问题。

在技术方面，小型模块化反应堆包含多种设计和技术，例如：可在近期部署的一体化压水反应堆；可在中期部署的、采用非水冷却剂/慢化剂的小型第四代反应堆；改装的或改进的紧凑环路式小型模块化反应堆（包括驳载浮动核电站和海床基反应堆）。以下三个重大进展提升了小型模块化反应堆的竞争力和吸引力：

（1）针对更小的反应堆使用非能动安全概念的可能性，它满足了日益苛刻的安全要求，同时又使得设计得以简化。

（2）工厂内模块化建造能力的形成，它应该可以降低总成本、缩短现场的工期。

（3）像"即插即用"概念一样，发电站完全在工厂内建造，然后运到现场接入电网。现场的主要操作是将发电站接入电网。

美国、俄罗斯、中国、韩国、日本、英国、阿根廷和法国正在研究的诸多小型模块化反应堆概念主要分为两大类型：陆基小型模块化反应堆和可运输小型模块化反应堆。

● 旨在实现核岛模块化的陆基小型模块化反应堆，需要安装在特定的位置，需配备土木工程及额外的附属设施、汽轮发电机组、电网接入等。

● 与运行厂址完全脱钩的可运输小型模块化反应堆，机动、灵活、换料维修后可重新部署，同时最大程度地减少新客户的整体采购时间。

作为典型的陆基小型模块化反应堆，ACP100（图 2）是由中国核工业集团公司开发的一个 125 MWe 小型模块化反应堆。ACP100 采用经实践验证的、实用的轻水反应堆技术。该技术有五大技术优势：一体化反应堆、固有安全特性、完全非能动安全系统、地下部署、双堆共享一个 250 MWe 的汽轮发电机组。

ACP100

图 2　中国核工业集团公司开发的小型模块化反应堆——ACP100

多个可运输小型模块化反应堆目前正在建设和开发过程中：

● 俄罗斯和中国计划在驳船上建设的小型模块化反应堆。俄罗斯的第一个机组正在进行安装收尾工作，中国正在加快 ACP100S（125 MWe，图 3）和 ACPR50S（50 MWe）的开发步伐。

● 法国正在研究的、能提供更多电力（160 MWe）的浸没式小型模块化反应堆 Seanergie（图 4）。该设计的陆上版本也在研究中。

图 3　中国核工业集团公司开发的 ACP100S 浮动核电站

图 4　法国 Seanergie 浸没式小型模块化反应堆

5.1.3　小型模块化反应堆所面临的挑战和解决方案

要让这些型号实现产业化，需要证明它们的竞争力、公众接受度以及国际原子能机构正在研究的移动式反应堆的监管框架。

要切实满足市场需求，新的小型模块化反应堆必须真正采用创新理念，绝对不能是目前的第三代反应堆的缩小版。

创新的设计可明显提升小型模块化反应堆在经济上的竞争力。与间歇性风电、太阳能发电、天然气发电和用于特定应用的柴油发电机相比，小型模块化反应堆是有竞争力的。

如果类似于"即插即用"、设计完全独立于安装地点的解决方案得到证实，那么它们可以成为满足市场需求并为能源转型做出贡献的最佳选择。

5.2　适用于大型反应堆的创新技术

商用核技术只有几十年的历史。由于安全要求很高，创新理论只能缓慢导入反应堆设计中。因此，无论是实施已经用于其他行业的技术，还是专门为核应用开发的技术，未来都有巨大的改进潜力。SMR 和大型商用轻水反应堆都将从这种技术中受益。核相关技术或从其他行业转让而来的技术包括：

1. 核能特殊技术发展

（1）高性能燃料：

● 燃耗增加，而膨胀和裂变气体释放量有限；

● 事故容错燃料，能承受高温不熔化，从而能在发生事故时防止或限制氢的产生。

（2）改进的堆芯内仪表，准确性更高，可减少设计分析和运行时的保守性。

（3）更加深刻理解堆芯熔融行为，优化事故工况下燃料在压力容器内滞留措施。

（4）实施最新模拟方法，实时耦合热工水力和中子计算方法，可大大提高设计和运行能力。

2. 从其他行业转让而来的技术

（1）核设施设计、采购、建设和项目管理的数字化（请参阅本报告第 8 章）。

（2）为低压回路系统采用新型复合材料以取代钢材。

（3）采用高机械性能和抗渗性能的先进混凝土。

应高度重视这些领域的研发工作；在不涉及知识产权问题时，应鼓励开展国际合作。

结论

核能是一项朝阳技术，有巨大的改进潜力（包括第四代反应堆以及小型模块化反应堆）。事故容错燃料等技术里程碑可提升安全水平、简化系统，从而增强所有轻水反应堆的竞争力。高度创新的 SMR 可以提供新的解决方案，进一步提高灵活性、推广分布式发电。它们还有助于新兴国家顺利地增加融资和在当地培养更多的核人才迅速进入核电行业。

第6章 放射性废物管理的现状与未来的展望

建议

　　放射性废物的安全管理是发展核电必须解决的一个关键问题。由于所有核电国家都面临这个问题，因此必须制定出相应的行政和技术手段来解决这个问题。处置最终的长寿命放射性废物需要克服许多重大障碍。深地质处置库是处置此类放射性废物的公认方法。中法三院建议加强各类放射性废物管理的科学技术研发活动，尤其注意对用作长寿命高放废物处置库的主岩的鉴定。中法三院还认为应加强与此相关的国际合作。

　　核燃料循环的各个环节都会产生放射性废物，这些废物的安全管理是证明核能可被安全妥善地管理的重要一步。在和平、可持续、可扩展地开发核能以及提升公众信心的过程中，放射性废物管理将起到重要作用。放射性废物可按其所含主要放射性核素的活度和半衰期进行分类。释热则是另一个分类准则。放射性废物的分类取决于各个国家的分级系统。由于需要保证慎密的辐射防护措施，且大多数情况下需排出余热，因此对长寿命中高放废物的管理是一项艰巨的长期任务。

6.1 燃料循环战略与放射性废物的特征

　　核燃料循环分为两类：一次通过式燃料循环（也称为开式燃料循环）和闭式燃料循环。在闭式燃料循环中，将会对乏燃料进行后处理，以回收利用铀（U）和钚

(Pu)。这些循环反映了将钚作为废物处置或用作核燃料的政治决定。在一次通过式燃料循环中，经长期临时储存，乏燃料子组件在衰变热得以衰减后，会被当成最终的高放废物予以处置。最终处置库是建在深地质岩层中的核设施。安全分析结果显示，主岩可隔离并包容放射性达数十万年。

在闭式燃料循环中，乏燃料会经过后处理，生成最终所谓的"处理过程中的"和"工艺性的"短寿命和长寿命放射性废物。"处理过程中的"长寿命放射性废物含有乏燃料中存在的所有放射性核素（铀和钚除外），具有高放射性。在开式燃料循环中，所有最初存储在核电站中的长寿命乏燃料放射性废物最终将被运送至深地质处置库。大多数利用核能的国家（包括法国和中国）目前已经采用闭式燃料循环。事实上，如果快中子反应堆中的铀和钚像法国和中国计划的那样进行多次循环利用的话，闭式燃料循环就将成为可持续发展的核能系统的基石。闭式燃料循环提高了天然易裂变资源的利用效率，降低了最终废物的毒性，因为它们不含钚（但必须在燃料循环中管理钚，直至循环利用结束为止），这在一定程度上减少了终极废物的量。

6.2 放射性废物管理的进展

放射性废物管理包括放射性废物生成后的控制、收集、筛选、处理等步骤。接着，废物经整备，装入相应的包装，以避免放射性核素的扩散。这些包装将被运输到存储地并最终得到处置。处置环节被视为核心目的和终极管理目标。接下来的几个章节将讨论放射性废物管理的现状（不含寿命极短的放射性废物或产生微量辐射剂量的放射性废物，这些情况是废物豁免和/或清洁解控所处理的主题，此处不予讨论）。

1. 极低放废物（VLLW）

无论极低放废物中含有的放射性核素是什么，它们的活度都很低，这种废物在地表装置中就可处置（浅填埋）。在核放射性废物中，它们的占比最大。事实上，核设施的拆除会产生大量极低放废物，在所有的退役放射性废物中占 50%~75%。因此，可以预测，未来将产生大量的极低放废物。除积极发展减容技术外，大型极低放废物处置设施的建设也势在必行。

2. 低中放废物（LILW）

如果只涉及短寿命放射性核素，那么通过建设和运行地表/次地表处置设施（包括浅壕沟、近地面混凝土结构、地下岩洞或隧道、大直径竖井）就可解决低中放废物的最终管理问题。有关低中放废物的整备、包装运输、现场接收、包装处置和安全评估的实践和经验都可从业界获得。

低放长寿命放射性废物的产量也很大，是一类难以管理的特殊废物。由于它们在很长的半衰期内均具有放射性，如果选择次地表处置，就必须证明长寿命放射性核素将长期与生物圈隔离。

3. 高放废物（HLW）

高放废物是含有大量的长寿命放射性核素有毒有害物质，如乏燃料的子组件或玻璃固化裂变产物包装和次锕系元素。高放废物必须储存冷却数十年时间，才能够转移到集中的深埋地质处置库。世界各国在选址和设计高放废物处置库方面做出了巨大的努力。比利时和法国选择了黏土层厂址，芬兰和瑞典将他们的高放废物处置库选在了相对均质的花岗岩中。德国、美国、英国和很多其他国家依然在适当的主岩中选址，目前为止已经排除了盐丘和火山岩。中国目前正在研究花岗岩和黏土岩，以确定高放废物处置库的选址。选址需要花费数十年时间，并需要在地下研究实验室进行测试，这似乎是不可避免的。截至目前，只有芬兰于 2015 年颁发了建设高放废物贮存库的许可。从目前的主岩选择研究工作和初步处置设计（包括处置库的密封）来看，可以预见高放废物将得到安全的隔离。

结论

放射性废物的国际管理有着明确的目标。这需要安全部署持续改进的解决方案，以管理不同种类的废物（极低放废物、低中放废物和高放废物）。这要求通过信息透明化解决相关冲突的经验，也需要增进公众对问题的理解以及提高公众参与度。对于放射性废物管理的未来发展，需要考虑以下几点。

1. 国家政策应始终关注放射性废物的管理问题

国家层面的决策是推动安全管理规划的主要力量，需依赖于强大的法律体系、成熟的科技研究和技术解决方案、详细的统筹规划和充足的资金投入。

2. 应加强对放射性废物管理科学技术的研发

为确保安全性，放射性废物管理需要先进的科学技术来处理和处置废物。深埋地质处置库的建设既构成核心问题，同时也是一个巨大的挑战。每个部件和控制系统必须在 100 多年的时间里保持运行，且处置库封闭时必须确保废物完全被隔离。加强基础科学领域的研发、鼓励技术创新、有效突破废物管理方面的主要瓶颈，这些都是至关重要的。

3. 应促进国际合作和创造广阔的未来前景

放射性废物的管理是需要核工业所有参与方持续关注的重要问题，必须促进国际合作，分享这一领域的知识、信息和技术。

附录 B

对每一个核国家而言，放射性废物管理的典型特征和趋势取决于诸多因素。主要的驱动因素在于该国选择的是开式还是闭式燃料循环。就闭式燃料循环国家而言，法国的放射性废物管理是一个很典型的例子。

今天，90% 的放射性废物是核电行业在制造、使用、循环和储存核燃料的设施的运行过程中产生的。从法国与乏核燃料有关的能源政策来看，预计这一数字不会改变。目前，法国有 58 座压水反应堆和 1 座欧洲压水堆系统，总装机容量 62 GWe，年发电量 420 TWh，所有核电站卸载下来的 UO_x 乏燃料子组件经过后处理，提取钚和铀，并在 MOX 燃料和 URE 燃料中实现一次循环利用。MOX 乏燃料子组件会作为钚的战略储备被存储，以备未来之需。反应堆的寿命期认为应达到 50 年。所有其他非核电乏燃料也会被后处理。

以废物的放射性水平——衰变期为判据，法国的放射性废物管理系统将废物分为五个族类，这反映了现实的管理观点，也与所谓的"放射性废物渠道"一致。表 1 显示了 2013 年放射性废物的生成量和反应堆寿命期 50 年内的预计总生成量。图 5 显示了法国近乎闭路的核电循环。在各个环节，圆圈的面积与 2013 年放射性废物的生成量成正比。

表 1　法国放射性废物生成量

放射性废物	缩写	到 2013 年的生成量/m³	放射性占比/%	预计总量/m³（长达 50 年寿命期）
长寿命高放废物	HL-LLW	3200	~98	10000
长寿命中放废物	IL-LLW	44000	~2	72000
长寿命低放废物	LL-LLW	91000	0.01	180000
短寿命低中放废物	LIL-SLW	880000	0.02	1900000
极低放废物	VLLW	440000	<0.000004	2200000
总计		1460000	100	4300000

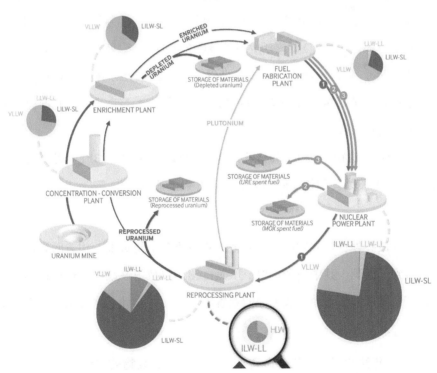

图 5　法国近乎闭路的核电循环（来源：法国国家放射性废物管理局）

1：铀氧化物燃料；2：混合氧化物燃料；3：循环铀。虚线：运行和拆除废物

核设施残留的和拆除的放射性废物也包含在总量当中。经过后处理后待处置的活性最强的放射性废物（5000 个长寿命高放废物包和 170000 个长寿命中放废物包）不

含钚或铀（Purex 流程中损失的除外）。此外，可以看出，相对于开式燃料循环而言，循环钍的闭式燃料循环使长寿命高放废物包装数减少约 90%（0.8 m³/TWh 比 8 m³/TWh），但长寿命中放废物的数量增加了（0.6 m³/TWh 比 0.08 m³/TWh）。

下面的章节将只关注长寿命废物和极低放废物的管理，它们是主要问题的根源。本章也给出了其他废物管理的一些指南：长寿命低放、短寿命中放、铀尾料和工艺增强的天然放射性物质（TENORM），这些问题已得到解决或正在解决的过程中。

1. 法国废物管理制度框架

法国国家放射性废物管理局（Andra）是负责放射性废物的核运营者。除其监管职责外，法国核安全局还为法国国家放射性废物管理局和法国原子能委员会、法国电力公司和阿海珐集团等放射性废物生产者提供安全指南。2006 年 6 月 28 日颁布的 2006－739 号法律针对各种放射性物质和放射性废物制定了管理政策，并①在考虑社会要求的前提下，设定了研究领域、里程碑和研究目标；②制订了管理计划（由法国核安全局和工业部详细拟定），包括放射性废物的临时存储/处置，以及长寿命放射性核素的分离/嬗变。它为全球的管理体系提供了一致的框架，并为放射性废物安全管理的持续改进做出了部署；③成立新的全国评估委员会（CNE2），负责评估放射性废物管理方面的科研进展。

2006 年的这部法律承接了 1991 年 12 月 31 日颁布的 91－1381 号法律。后者规范了分离/嬗变、地质处置和长期储存方面将要开展的研究工作。

2. 进展与挑战

从整备到最终处置，放射性废物管理一直在不断取得进展。相关问题也通过研发长期规划得到了处理。其中一个重要成就是法国国家放射性废物管理局选择了黏土地层作为长寿命高放和长寿命中放废物的处置库厂址，并设计了深地质处置库 Cigéo。

3. Cigéo 处置库（长寿命高放、长寿命中放废物）

上述 2006 年法律规定，长寿命高放和长寿命中放废物处置设施应至少在一个世纪里是"可逆"的。这里考虑的可逆性，是指后代继续建设、进一步利用后续存储步骤，或重新评估之前所做的选择、改变管理计划的能力。这包括按照与运行和储存终止战略一致的方式将已储存了一段时间的废物包回收的可能性。

名为 Callovo-Oxfordian 的黏土层已被选为主基岩。Cigéo 的建设将很快在 Bure

启动。Bure 位于地下实验室 LMHM（Laboratoire de Meuse Haute-Marne）附近，该实验室已由法国国家放射性废物管理局运行了超过 15 年。这个 130 m 厚、500 m 深的黏土层位于巴黎盆地东部，具有很好的横向连续性，在成分和结构上很均匀。对该地区区域和局部地质、水文和地球化学进行的研究发现，穿过黏土的上下含水层之间没有断层缺陷或与对流的联系。黏土孔隙水有很长的停留时间。LMHM 在当地已测定了水的成分以及扩散速度最快的放射性核素的扩散系数。数据显示，黏土可在长达 100 万年时间内提供阻止放射性核素迁移的有效最终屏障。其机械性能允许修建耐久的功能性地下设施，可在特定房间分别处置长寿命高放和长寿命中放废物，并保证处置库能够持续运行长达一个多世纪。Cigéo 的现场勘查和设计工作始于 1991 年。

4. 长寿命低放废物

长寿命低放废物的组成有含镭及其放射性衰变产物材料、含氯 36 和碳 14 的大小石墨块、含有微量锕系元素的沥青液滴包。它们被储存在不同的地方等待处置。法国国家放射性废物管理局正在黏土层中寻找专门的次地表场址。在这种条件的地点处置废物存在挑战，因为需要包容的长寿命放射性核素与生物圈非常接近。对长寿命低放废物进行工艺处理也许有助于降低处置的难度，目前正在进行相关研发工作。长寿命低放废物的量预计将达到 200000 m^3。

5. 短寿命低中放废物

短寿命中放废物（$1.9 \times 10^6 \ m^3$）主要在发电过程中产生，在当前和可预见的未来，此类废物都在位于苏莱内迪伊（Soulaines-Dhuys）的奥布储存中心（Centre de Stockage de l'Aube，CSA）进行包装和处置。CSA 的处理能力为 $10^6 \ m^3$，该中心于 1992 年开始投入运行，预计将运行到 2100 年。上一个处置库英吉利海峡储存中心（Centre de Stockage de la Manche，CSM）位于迪居勒维尔（Digulleville），其处理能力为 $0.5 \times 10^6 \ m^3$，该中心于 1969 年开始投入运行，自 1993 年封闭以来一直处于监控状态。

6. 长寿命极低放废物

长寿命极低放废物目前在特定中心进行处置：位于莫尔维利耶尔（Morvilliers）的重组和储存工业中心（Centre Industriel de Regroupement et de Stockage，Cires）。其核准处理能力为 650000 m^3，可扩展到 900000 m^3，预计到 2030 年满容。除对

Cires 进行扩容外，还需开设另一个大容量的处置中心，以处置拆除过程中产生的长寿命极低放废物。

由于法国的放射性废物管理体系没有定义清洁解控水平，而长寿命极低放废物的预期产量巨大（估计可达 2.2×10^6 m³），因此很可能会导致处置困难。多年来，研究组织、行业和政府监管机构一直在开发创新方法管理拆除时所产生的、被归类为废物的材料（虽然它们不含外加放射性或含量很低）。为了做到这一点，必须开发批量测量材料极低放射性活度的方法，以支持创新型极低放废物管理战略。法国国家放射性废物管理局和废物生产企业首先正在考察各种显著减少极低放废物产量的可能性（例如：通过大量循环利用金属部件）。然而，循环利用也存在诸多困难，例如：为实现再利用，起码需要定义去污过程中应达到的各项阈值。

7. 工艺增强的天然放射性物质（TENORM）

到目前为止，非核工业产生的 TENORM 都在配有固定阈值放射性检测设备的非核工业废物技术中心进行管理。这种管理方式正由于 TENORM 的量和活性受到重新审视。管理此类物质有与低放废物类似的困难。

8. 其他废物

法国有多处已开发的铀矿。大约有 50 Mt 铀放射性废物（尾料）在其生产地附近的 17 个场地就地处置，这些场地位于废旧露天矿场、盆地、堤坝围绕的谷底中。这些场地不仅受到监控，而且也须遵守特定的监管条例，以便对氡的迁移及放射性衰变产物进行管理。加工黄饼产生的大量放射性废物（与采矿放射性废物有点类似）将被储存在 Malvesi。

总之，依照后处理/循环利用所有乏燃料的国家战略，法国在其核安全局的管理下建立起了安全的放射性废物管理体系，以便为其核活动提供支持。深地质处置项目 Cigéo 将在几年后启动。极低放废物管理的现阶段目标在于减少拆除反应堆和核设施产生废物的体积，并选定新的大容量处置场。包括公众在内的所有利益相关方进行了多次讨论，结果逐渐清晰：人们想要的是安全，而非"安抚"。关键问题在于：人们如何才能对寿命如此长的放射性废物的管理感到放心。相关法律规定了放射性废物处置的可逆性，解决了伦理和将职责传递给后代的问题，有助于公众认可的实现。为落实处置可逆性，下一步工作在于建立管理制度，依法保证公众参与。

第 7 章　安全领域的技术支持机构

建议

● 监管部门的专业技能（或围绕自身建立或由技术支持机构（TSO）提供）是确保高水平核安全的根本。

● 应尽最大可能维持 TSO 的独立性和透明性，这一点决不妥协。

● 应持续提高 TSO 的能力，以让它们有能力评估新颖而多样化的创新技术（包括第四代反应堆）。

● 应加强国际 TSO 间的合作，以更好地统一要求、评估方法和准则。

● 核发达国家的 TSO 应协助和支持新兴国家发展本国的安全技能。

所有核能相关的活动均受到深入的监管（参见国际原子能机构的 No. GSR《一般安全要求》第 1 部分）（第 1 版）。

核设施的整个寿期中，业主/运营者对安全负有首要责任，而监管机构应完全保有监管业主/运营者的权力。

除为具体设施颁发施工许可证、初始装料许可证、运行许可证或退役许可证等行政和法律任务外，监管机构需要强大的技术力量/组织提供支持，以制定合理的安全规则、条例和要求，并评估被许可方是否合规。

监管机构可以采取多种组织形式获取核专业能力，这种能力可由监管机构在内部建立，也可承包给第三方。当核安全专业技术能力进行分包时，这个优选的专业组织通常会被称为技术支持机构（TSO，也被称为技术与科学支持组织）。

在美国，美国核管理委员会拥有自己的专业力量，制定核安全法规和管理导则，并按照极其规范的要求对可接受性进行评估。在法国，监管机构（法国核安全局，

Autorité de Sûreté Nucléaire，简称 ASN）负责制定通用导则，并在一、二回路的完整性领域拥有强大的技术力量，以与高校和研究机构签订的合同作为补充，但在特定的核工程方面依赖于法国辐射防护和核安全研究院（Institut de Radioprotection et Sûreté Nucléaire，简称 IRSN）的专业能力。IRSN 是从法国原子能委员会前安全部分拆出的一个机构，后与卫生部的放射物理部合并。基本上所有的技术安全评估工作以及其他一些任务均由 ASN 指派给 IRSN（也就是法国的 TSO）。

在中国，监管机构（国家核安全局，简称 NNSA）负责发布核安全条例和导则、起草和颁布核安全法规、监督其实施、制定核安全原则和政策。所有民用核设施均受中国国家核安全局监管。中国原子能机构（CAEA）负责全国的核应急事务。

中国的大部分核项目由国家核安全局（NNSA）通过多个 TSO 进行评估：①环保部的核与辐射安全中心（NSC），提供 NNSA 所需的大部分技术支持工作；②环保部的辐射监测技术中心（RMTC），其主要职责是在全国提供辐射监测支持；③苏州核与辐射安全中心（SZNSC），在质量保证、机械设备、研究堆领域提供技术支持；④核设备安全与可靠性中心（NESRC），参与审查核承压部件；⑤北京核安全审查中心（BRCNS），它是在北京核工程研究院（BINE）中设立的首个 TSO，其主要职责是为非由北京核工程研究院设计的核工程项目提供技术支持。所有安全评估工作均外包给这些 TSO，特定的领域承包给了高校和研究机构（如清华大学、中国原子能科学研究院、中国辐射防护研究院等）。

审视全球在该领域的经验，我们可以发现，没有证据表明某种组织模式优于其他模式。很多模式都是可行的，只要它们：①适应各个国家的技术传统和管理体制；②将最佳技术专长提供给核监管机构；③促进这些技术专长的发展。

然而，一些普遍适用于核安全技术专长的基本原则还是可以被归纳出来的，尤其是：

（1）独立性：具有技术专长的主体（TSO）不应承担可能影响其中立性或可能导致评估自身的工作的任务。如果该主体为多个客户（包括安全监管机构）服务，则应识别并防止潜在的利益冲突。

（2）透明度：TSO 的组织应有利于巩固其运营的独立性和透明度，包括全面披露该组织的报告（除非涉及重大的保密问题）。

（3）质量保证：从接受任务到项目交付，专业机构都应严格按质保程序行事，确保对其专家的能力、其技术的可追溯性进行评估。

技术专长主体可提供培训服务，管理自己的研发项目，这些都是提高员工素质的手段。

TSO 不应闭门造车。虽然核安全对所有国家来说都属于主权问题，但经验和最佳实践是应被共享的。业主/运营者每天都在世界核能协会的框架下分享它们的经验，并定期组织同行评估。在核安全公约规定的工作组或同行评估的框架下，国际原子能机构和经合组织为安全监管当局搭建了多个交流与合作论坛。TSO 没有这样的论坛；然而，世界上主要的 TSO 均主动在国际原子能机构的框架下定期召开会议。2014 年第三届 TSO 国际会议在北京召开。这样的交流是值得鼓励的。虽然各个国家的安全要求不尽相同，但在 TSO 之间分享经验和技术只会让大家受益。

新兴国家的核能发展遇到了获取核安全专业技能的问题。当引进核电技术时，通常会考虑"参考电站"。只要满足独立性和透明度的要求，客户理应从该技术的来源国获取安全技术支持和参考电站的安全案例。此外，更先进国家的 TSO 应尽量帮助和支持新兴核国家建立并发展他们自己的安全技术能力。

回顾 50 多年的商用核电技术发展史，TSO 发挥的重要作用不容低估。核事件或事故分析表明，缺乏独立透明的 TSO 以及不能实施严格的质量保证计划，已经造成严重的后果。将来，TSO 应一如既往地服务于全球核工业。未来面临的一些关键挑战包括：

- 保持并提升 TSO 的技术专长，使之有能力评估多样化的创新技术（包括第四代反应堆）。
- 加强国际合作，以更好地统一要求、评估方法和准则。
- 支持新兴国家的安全监管部门和 TSO 建立自身的专业技术能力。

结论

核安全监管部门需要依赖强大的专业技术能力，这种能力可由监管部门在内部建立，也可承包给 TSO。这两种组织模式并无优劣之分。

TSO 应严格遵守独立性、中立性、透明性和质量方面的规定。虽然 TSO 在很大程度上能受益于国际合作和经验交流，但目前还没有专门的合作论坛和同行评估。然而，先进核国家的 TSO 能够并且应当大力帮助后来者发展他们自己的核安全专业技术能力。

第8章 未来面临的数字化和新设计方法等方面的挑战

建议

核工业应在所有开发阶段都充分发挥数字化的优势。

● 利用信息技术的进步（包括计算机速度和存储能力的指数增长），通过模拟反应堆正常和事故工况运行，实施更精确的设计标准和规范。为此需与安全监管部门积极互动，以便为新技术方法获得许可铺平道路。

● 仪表和控制系统已经实现数字化；但监管部门之间的安全要求尚未统一，导致同样的功能可以用不同解决方案实现。应对网络安全等新威胁的法律法规也未统一，而这方面的进展需在国际层面上取得。

● 许多行业都能充分利用新数字技术进行设计和项目管理（CAD 工具，项目生命周期管理）。核工业也能从这些工具中获益良多，应鼓励借鉴该领域的经验。

在生产低碳廉价电力的同时，核工业必须在不影响安全的前提下减少成本、缩短新核电站的建设工期、更好地控制现有电站的停机时间。

要应对这些挑战，需要充分利用数字化革命带来的所有新工具，尤其是"工业4.0"、产品生命周期管理（PLM）、3D 扫描生成现有设施数字模型、大数据和物联网、虚拟现实和增强现实、3D 打印技术等。

此外，核能在开放的全球经济中发展，在中国体现为聚力"一带一路"，在欧洲则体现为开放市场竞争。

核工程的数字化革命将有助于增强竞争力和经济效益，同时提高安全性；工作组织也可通过数字化变革显著提高效率和收益，因此必须适应数字化。

很多核工业的重要议题都与数字化技术直接相关：

- 更好控制新核电站的施工周期；
- 通过更好地规划、安排和控制停机，提高现有电站的利用率；
- 在役机组寿命的延长；
- 第三代核电站竞争力的提升。

在役机组在设计阶段没有享受到数字模型带来的更多益处，可在运行过程中进行更新，通过 3D 扫描创建数字模型与分包商共享，这样可以缩短维护操作的时间、提高维护中试验的可操作性、使操作员培训更加简便，直到电站退役为止。数字模型还可提供相关工具，以最大限度减少工人暴露在辐射中的时间。

对新项目而言，考虑到核项目的生命周期较长（一般是 60 年），且安全要求极端严苛，使用数据管理工具实时管理反应堆配置（涵盖规格、设计、施工、运行和拆除等信息）是至关重要的。

此类工具（例如：PLM）目前已经存在，且将在航空航天领域得到长期应用。

未对数据进行完全监控是施工阶段出现延迟、因返工导致成本超支的主要原因。

要成功部署产品生命周期管理数字平台，就必须改变行业的组织模式。这要求：

- 系统工程及"建筑工程师"（"总工程师"）对整个系统（核岛、常规岛、泵站……）的设计、施工到试运行阶段负全部责任。

- 严格的数据管理，包括必要时授权变更配置的程序；使得数据不再分散在纸质文件上，而是嵌入到一个单一的 PLM 数字平台中，以供项目各方使用。

PLM 使检查施工前设计完整性、施工或维护次序的交互式数值模拟也变成了可能，以便评估可达性和优化施工方法。

项目的主要利益相关方应接驳至 PLM 数字平台（土木工程、常规岛、核岛、分包商等），以搭建统一的数字管理系统，提供创新设计方法，并大大缩短项目工期。

制造链的数字化（尤其是无纸化）还将改进数据的可追溯性、部件规格书和数据表的存档和检索，用单一数据源全局改善质量保证。

在核项目的整个生命周期中，智能设计方法对改进核工程设计起到了至关重要的作用。

概念设计阶段必须做出决定项目构架的关键决策,包括核电站输出、热阱布置、厂区总图设计以及建筑物的主特征(结构、大开孔等)。上述因素必须共同优化,而通常优化是依次进行的。设计过程中应尽早使用三维 CAD 工具,以加速替代优化和评估解决方案······这种逐步优化的无缝衔接需要高水平数字化技术(包括静态和动态计算)。经验表明,设计的简单透明无论从哪个角度都是概念设计阶段追求的主要目标,且进步空间依然存在;软件设计者应考虑这些设计前期所需的特定功能。

初步设计阶段,依托创新数值模拟手段可实现更好的优化,从而提升竞争力。合格可靠且经过实验验证的仿真模型,既可用于引导设计的优化过程,又可提供设计反馈。仿真技术仍然需得到核安全当局许可,但核安全当局有时过于严苛,反而对创新造成阻碍。

虚拟现实/增强现实技术旨在将三维设计模型集成到统一的软件环境中,并将不同部件的几何特征还原到全尺寸数字化 3D 模型中。利用这些技术可以更有效地优化反应堆和辅助系统的布置以及与土木工程之间的接口。从设计到试运行,核电站项目整个生命周期的产业绩效都可以通过数据发掘和估值技术得到提高,正如其他工业领域中所认识到的那样:通过大数据技术分析整理从核电站中收集的大量数据,可使预测性维修更准确高效。

核项目意味着海量数据管理,其数量远多于汽车或飞机行业目前的水平。然而,事实证明,类似的 PLM 方法结合 CAD 工具,对于提高数据质量和简化项目管理非常有效。这些方法和工具的应用可增进项目的可追溯性,对质量保证和安全也有益。重大益处还有很多,要实现它们还需得到包括安全监管部门在内的所有利益相关方的配合。

此外,数字化仿真还应用于培训用全尺寸模拟机,以改善应急准备和提升严重事故响应能力。

25 年来,核电站的仪控系统大量运用了数字技术。然而,核工业应用数字技术仍需克服诸多挑战(包括对软硬件进行鉴定以使其满足安全要求)。科技成果的产业化是核能行业提升竞争力的关键,但这需要全行业和监管机构的参与。适用于安全相关关键软件的法规标准仍有待制定。此外,国际法律法规在阻止、监测针对核电站的恶意网络攻击等事件方面还存在大量空白。因此,我们应加强对某些立法的制定,同时还应关注实体隔离、防火墙和有效管理措施等相关安防技术。

结论

从概念设计到施工、运行和维护，直至反应堆退役，无论在哪个阶段，核工业都能受益于全面数字化技术。产品生命周期管理（PLM）正是这样一种工具，并在航天业已有广泛而成功的应用。其他工具还有数值模拟、三维 CAD 和虚拟/增强现实。以上工具的利用可大幅提升工程效率，避免工程延期和成本增加，同时也将提高竞争力。

在设计阶段使用仿真工具（中子、热工水力等）大有前景。多年来，核电站一直在使用数字仪表和系统。在以上两个领域，世界各国的安全监管部门需协商一个统一的审查方法。

在任何情况下都必须高度重视网络安全，以便及时发现并防范网络攻击。

第9章 核研究设施和基础设施的重要性

建议

对于（无论是核能大国还是新兴核能国家的）核研究组织的任务，可以提出如下两点建议：

● 在全球化背景下，考虑到核研究设施不断增加的成本及（至少部分设施）相对较低的运行率，建议探索互助共建并共同使用大型和/或重型和/或特定核设施的可能性，包括研究堆、热室、辐照考验材料实验室、仿真与计算中心等。该建议对现有设施也有价值，但尤其适用于新建设施。

● 为使新兴核能国家能够以可接受的成本使用研发设施，我们鼓励那些核工业发达国家的研发机构和设备制造商提供价格低廉的实验设施，以满足核领域（也包括前端和后端活动）的实验研究的特定及一般要求。

在几乎所有工业化国家中，核基础设施都源于国家原子能委员会，这些原子能委员会都具有浓厚的学术研究气氛。这些机构不仅拥有核研究和教育中心，而且还是大多数国家的核安全机构、安全技术支持机构（TSO）、废物管理机构的摇篮，同时也是核工业的重要组成部分。

出于保密和防止核扩散的考虑，军用设施的项目计划、基础设施和人力资源很早就与民用设施分离。值得注意的是，民用核研究在很大程度上一直是公开的，始终遵循国际原子能机构（IAEA）的透明化原则，对国际化社会公开。

从一开始，核反应堆行业和燃料循环行业相互独立。前者很快引入竞争性市场机制，其中一些市场参与者是从国家能源委员会剥离出来，而其他参与者（如西屋电气）的崛起与政府机构无关，他们在反应堆盈利的庞大现金流的支持下，完成核电站的设计、规划和建造。西屋电气在核领域所取得的辉煌成就对法国、日本、韩国和中国的核能产业产生了强烈的影响。在第二代和第三代反应堆市场中，核研究中心通常（但并非总是）局限于特定项目支持（特别是对于供应商或电力公司在反应堆运营中面临的高难度问题）。

今天，这些研究中心仍在继续核反应堆燃料性能的研究，国家实验室也在进行反应堆延寿的研究。几乎所有情况都需要使用设在核研究组织（研究堆、实验设施等）内的具体设施。在法国，研究设施为法国电力公司（EDF）和法国原子能委员会（CEA）共有。

由此可以得出以下结论：燃料循环产业与核研究组织的关系更为密切。该领域的下游企业通常开始是从这些研究机构中剥离出来，成为独立企业（尽管创办研究机构持有大量股份），然而两者仍会开展合作研究项目并进一步发挥共同创新能力（后处理或废物管理等）。另外，燃料循环产业的浓缩业务主要由私营企业自行研究。

过去 15 年来，有几个问题对核研究组织产生了巨大影响，在下列情况下，必须重新定义它们的角色或至少它们的运营方式：

- 先进第二代反应堆的延寿；
- 第三代反应堆的问世；
- 第四代反应堆的研发；
- 新兴国家发展核电，或发展核电的意愿；
- 严重事故；
- 全球各地核电成本的增加，主要是提高安全性带来的后果；
- 一些工业化大国逐步"淘汰核电"；
- 用训练有素的核电人力替代"第二代员工"的巨大需求，以及（至少在西方国家）年轻一代对科技职位相对不满（参见第 10 章）；
- 退役项目及相关反馈的大量增加；
- 最后，也是非常重要的一点：小型模块化核反应堆模式的问世。

在没有明确宣布的私人融资的情况下，安置和管理创新型核项目或规划的一种较好的解决方案，是依托现有的核研究组织。这些组织拥有经验丰富的团队，既能进行短期研究，同时为选择或建立工业实体留出时间，又能承担燃料循环等方面的中长期研究（例如：第四代反应堆、小型模块化反应堆等）。

然而，在很多新兴国家，研发机构主要由高校、教授和学术人员进行管理。尽管这为开展研究活动奠定了基础，但是它还应得到大量工业或起码是技术资源的补充，从而在项目启动时为参与各方提供技术专长和附加值。

不过，这并不妨碍主要来自高校和学术研究的颠覆性创新的应用。在学术团队和行业公司之间建立高效的合作关系总是大有裨益的。

结论

即使是决定完全借助外力（例如"建设-拥有-转让（BOT）"或"建设-拥有-运行（BOO）"模式的设施）的新兴核国家，也应当具备最低限度的技术能力来管理与电力公司之间的合作，履行自己的责任，确保运营期间的人员安全和全球核社区安全。"所属国家业主的工程师"在安全方面的职责不可分包。不过，国外安全技术支持机构（TSO）的支持是有益的，他们在其他国家取得的经验值得借鉴。

在这种框架下，可以在国际原子能机构定义的核能计划实施组织（NEPIO）的职责的基础上增加研发任务，从而对 NEPIO 或其他国家机构实施的建设项目提供支持。

第10章 教育与培训方面的挑战

建议

中法三院认为，必须加强教育以满足核行业对素质过硬的工程师和技工的需求，且应增强核工业的吸引力，方式可以是促进科学技术、工业和高校之间的互动、创造受欢迎的职业前景。建议如下：

● 改进教学方法，在面向工程师和技工的课程中整合现代 IT 技术（例如：模拟机、多媒体、网络学习和虚拟现实），另外还要使课程更多样化，以包含 PLM 等更广泛的技能以及行业文化的继承；

● 提高核电站的成本效益和安全性，将数字化在施工、运行和维护中的使用整合到学生课程中；

● 充分借鉴现有经验教训，为第三国提供更好的培训服务，启动中法两国在该领域的合作。

过去十年，教育和培训背景发生了很大的变化。

(1) 在 2020 年至 2030 年期间毕业的工程师、科学家和技术人员将一直工作到 2065 年和 2075 年。核工业从业人员必须：

● 运行、维护已安装核电机组，并对其进行现代化改造，同时拆除一部分核电机组；

● 建设新的"第二代、二代＋、第三代"核电站，然后运行和维护这些核电站；

● 设计新型核电站（包括第四代堆、小型模块化反应堆……）；

● 开发先进燃料循环（包括废物管理）。

在法国，虽然今天核工程师在核行业中所占比例高于 20 年前，但仍对技工有需求（技工：通过职业高考的高中毕业生，高中毕业后接受 2～3 年的专业培训和教育的从业者）。技工占从业人员的一半左右。

中国则在过去 30 多年来一直在建设核电站。尤其在 2005 年启动了一批新核电站的建设之后，中国越来越多的高校设立了核技术课程，为这个行业培养了合格的高素质人才。但是，目前，他们的专业技能还不够强——雇主必须系统性地长期培训他们的员工，提供在职实践机会，让他们尽快上手，高效履行各自的责任。

（2）核工业的发展环境与 20 世纪 80 年代相比已大不相同。核电行业目前正处在十字路口。第一，全球已经对应对气候变化的必要性达成共识。绿色能源和低碳过渡政策为中国核电工业的发展以及为中法两国在第三国合作开发核电市场提供了更多的机会。

第二，随着风能和太阳能技术的迅猛发展，其运营成本急剧下降。然而，在福岛核事故后核电的安全要求变得更加严格，新核电站项目和现有核电站的安全改进所需的投资大幅增加。因此，核电原先相对有利的竞争优势正在逐渐减弱。

第三，公众对核安全更加担忧，核工业的继续发展面临挑战，例如：如何与利益相关方沟通并赢得他们的支持。

第四，由于核电发电能力趋于不稳定以及电力市场的放松管制，因此核电站的持续盈利将变得不确定。

（3）核工业的安全问题变得至关重要，因而保守主义抬头，遏制了创新。比如，核工业的数字化远比航空业等其他行业滞后。但随着互联网、大数据、人工智能和其他技术的发展，更多的数字化技术、智能技术将被运用到数据管理、产品生命周期管理（PLM）、数字模拟器（DCS）等领域。这些领域所需的知识必须传授给核工业从业人员和政府监管人员。这是核工业可持续发展的先决条件。

（4）由于核安全问题极其复杂，核工业的所有从业人员必须具备丰富的专业知识、技能、经验和工程文化背景。因此，必须为新员工制定适合的知识管理和人员培训体系，提供全面的培训，培养胜任的教师，促进更高效的制度化跨代合作。对各个层级的所有员工都应如此。

中法两国的核工业和"教育生态系统"（高校、工程院校、研究机构、技术研究所）将通过合作应对以上情况：

● 教育生态系统有责任重新调整学生的课程，以便将产品生命周期管理、环境问题、跨学科和系统性问题、现代仪表等纳入其中，而不仅仅像过去那样只有基础科学

（如中子学、热工水力学、材料力学、结构力学等）和/或基础工程和传统项目管理。

● 核工业有责任雇佣背景更多元化的员工，从而考虑新理念和"外来的"理念，积极、稳定地促进创新的工程手段。

● 为强化核安全文化、全面的质量管理，应加强相关核工业领域的人员培训合作，应共享培训资源，同时还应统一培训标准。

● 应创造条件，让核工业的科学家和工程师有机会在教育界兼职或工作一定年限，同样也让教授/研究人员在核工业兼职。

这里应该提到的是，中国和法国就核工业人员培训发起了诸多国际合作项目，如刚成立不久的中法核工程与技术学院（简称 IFCEN，隶属于中山大学）的成功经验。该学院中法团队合作设置并教授创新课程，2016 年 6 月，首批 80 位工程师在这里完成了学业。

在国际原子能机构的支持下，中国政府继续在哈尔滨工业大学为新兴国家培养核专业硕士和博士。为支持清华大学针对新兴国家学生开展的核工业培训项目，中国教育部每年将向 30 位留学生颁发奖学金。针对核工业开展的定向培训在中国的高校得以普及，加速了应用核电技术人才的培训。校企之间双赢的培训合作，使毕业生获得了就业机会，企业招到了训练有素的人才。培训合作使毕业生的技能培训得以改进，加快了他们的成长，同时也提高了教授课程的水平。

最重要的是，核工业必须预估人力需求，并看清目前工业人员的年龄金字塔。这一艰巨任务 15 年前在法国已经发生过一次，因为此前预计第二次世界大战后的婴儿潮一代将在 2000 年左右退休，当时核工业必须未雨绸缪。如今，法国则需准备在国内大部分核电站短期加速建设过程中大量雇佣人员的退休潮。

为不同种类的岗位培训公司各个层级现有的员工时，我们不仅必须考虑系统的更新或完全替代、新软件的出现——情况一直如此，而且还要考虑为人员提供不同的培训。比如，核工业新一代从业人员的思维方式、价值观和学习习惯全都在改变。IT界的不断发展让我们得以利用多媒体、视频、在线学习、虚拟现实技术来培养各个层级的员工——不仅仅是工作人员和工程师！因此，核工业监管机构应适应、支持和推广将在核工业中应用的新培训方法。

比如，模拟器在很早以前就开始被用于电站操作员的培训。随着建模、仿真和计算能力的进步，模拟器变得越发强大（尤其是在模拟事故后工况方面），能对电站运行提供更好的培训条件。此外，仿真、虚拟现实等技术领域的进步开创了以不同方式高效培训维护操作员的可能性（尤其是对于那些必须在辐射条件限制了干预时间的区域进行作业的人员）。同样，其他行业的做法可能适用且有助于核工业的各种操作员。中国广核集团的培训中心创新了多种模拟培训课程和工具，将学习、教学、培训和研究集成到一起，并建设了多级培训体系和设施。他们开发出独立的便携式模拟器。"门对门培训"和"边学边练"已经变成现实。这些创新手段改进了培训的质量。

要想取得成功，核工业还必须留住现有人才，吸引新人才。事实上，核工业不再被视为科技含量最高、众人追求的行业。关键是更新与科技界的互动，一方面通过对现有人员进行新式培训，另一方面通过对未来员工进行教育。

对于想要开发利用核电站技术的国家，应鼓励核工业人员培训方面的合作，尤其是要为国际项目发现和培养人才，以促进全球核电规划的发展。此外，应加快国际核工业标准、地方法规和技术标准的制定和实施，以及对核电站设计、施工、运行技术知识和能力的开发与运用。由于中法两国已建立成熟的核专业教学体系，拥有良好的国际专业培训基础，因此两国希望在这一领域开展合作，以便为第三国提供更好的培训服务。

结论

中法两国都需要对更多核工程师和技工开展更好的教育和培训。而且，在20～25年后，法国的核工业将出现第二轮退休大潮，同时也会带来招聘需求的加速。对此，两国应该未雨绸缪。重新加强与教学科研机构之间的互动，有助于吸引新人才。教授/研究员将有机会到企业中工作，而兼职教育工作的企业科学家和工程师则能丰富培训内容。培训项目也可利用现代 IT 技术的优势，通过模拟和其他教学方法，实现更经济有效的培训。此外，培训项目应涵盖更广泛、更多样的技能和能力。相比风能和太阳能技术的快速发展和成本下降，核工业却因为更严格的安全要求而遭受成本上升的挑战。要想解决这种成本上升问题，当今的学生必须熟悉核电站建设、运行与维护的所有阶段和环节的数字化技术。

第 11 章 核电项目工程设计与管理

建议

● 核电项目非常复杂。中法三院建议使用最先进的现代化工具来降低这种复杂性。

● 在整个项目实施过程中都应进行风险评估。

● 由于项目实施阶段安全规则的变化会给工程设计带来很大的不确定性，因此在启动一个项目之前，应充分论证和固化这些规则。

● 恰当地使用先进的管理工具可应对此类大型项目的工程挑战。

核电项目的规模和复杂性都是特有的：在不计算子部件的情况下，需要设计、制造、安装、测验和调试的对象仍有数十万之多，通常比航空或汽车项目高一到两个量级。此外，还需要符合复杂的安全要求和质量等级标准。由于这些原因，核电项目周期通常长达 5 年或更久，这还不包括场地准备、长周期物项采购以及全部基础设计和大部分详细设计开发。因此，核电项目控制是一项具有挑战性的工作，控制不力可能会导致严重的后果。比如，在三里岛核事故（1979 年）发生之后出台了新的安全要求。在等待新安全要求生效的这段时间，在建核电项目都被暂停。因此，许多项目周期延长了一倍，并且还有许多项目被取消。最近，西方有些核电项目因开工之前的设计不够完整以及供应链环节的经验欠缺等原因而陷入困境。另一方面，中国以及其他亚洲国家的许多项目表明核电项目的复杂性可以被克服，通过恰当的设计和利用相应的项目管理方法可以很好地控制核电项目的实施。

对于新型首个反应堆（FOAK）项目而言，其风险更大。对于这些项目而言，设计阶段的研发和技术验证是控制风险的关键所在。一般而言，采取有效措施确保研发成果在设计阶段可用是最为重要的。因此，新技术的研发过程需要与项目的进度计划紧密结合。此外，FOAK 项目的获批通常需要较长时间，因为监管机构要求提供新功能有效性的证据。因此，FOAK 项目需要获得更多关注，在新概念通过验证并获得许可之前不应开工建设。

当核电项目基于一个参考电站时，在实施阶段面临的主要困难和风险将涉及以下方面：设计质量和稳定性、供应链性能、部件的按时交付以及现场所有活动的协调。

最新的项目经验表明，至少 70% 的详细设计必须在开工建设之前完成，这要求提前进行采购活动，以便获得进行静态和动态分析、锚件确定和施工板材开孔等方面的相关数据。只有对土建工程、管道和电气布局等没有影响的小开口项才可留待以后解决。必须利用相同的 CAD 工具和数据库，在建筑设计工程师以及分包商和供应商之间完全无缝地分享设计（参见第 8 章）。许可要求的稳定性是前提条件，监管机构应理解"建造和运营联合"许可能够带来更好的安全性和质量。

采购活动和供应链是核电项目获得成功的另一个关键因素。中国和法国通常采取业主所有模式。在这种模式下，业主拥有自己的设计和项目管理能力。这种模式被证明可以获得成功，其前提是采用最新的技能和方法，并借助市场上最好的工具和经验。

设计、采购、建造和调试阶段的协调工作需要先进的项目管理工具，这需要将部件数据库、进度管理工具、文档控制系统等融合在一起。这样的设计工具现已存在（例如：项目生命周期管理），同时必须制定和实施严格的程序以便充分利用这些工具的强大功能（参见第 8 章）。

最后，还需要有涵盖质量、安全、进度和成本的强有力控制。这些控制举措需要尽可能独立于项目管理，并可直接向最高层汇报。

为了确保核电项目质量和进度，需要采取一些必要的措施来降低建设成本，同时加强成本控制。这可降低核电项目成本，提高成本效益。成本管理可受益于挣值评估。挣值管理可预测进度和成本偏差，并提供确保实现项目目标的措施建议。

附录 C 提供了关于项目风险的详细说明。良好的设计和项目管理可更好地控制和消除这些风险。

结论

鉴于核项目的复杂性，因此需要在设计质量、供应链可靠、使用现代 CAD 和 PLM 工具控制项目的一致性和进度方面给予最大的关注。

近期的经验表明：核电项目如果能得到有效控制，那么就可以在预算内按时交付核电项目，同时满足严格的安全和质量要求。

在其他行业经常应用的管理方法可以给核工业提供很大帮助。

附录 C　核电项目建设面临的主要风险

1. 融资风险

核电项目属于固定资产投资，稳定的现金流是项目实施能够成功的前提条件。项目的债务资本与权益资本的比例取决于资金成本与可接受风险之间的平衡。一般而言，项目融资风险涵盖了项目的全部风险。具体而言，重大项目融资风险包括偿债能力、投资能力、再融资和财务风险（如利息风险、汇率风险等）。

项目融资风险还在很大程度上受核电项目建设周期的影响。项目延期不仅会导致成本增加，在首期贷款到期电厂没有做好投运准备的情况下还会导致一系列法律问题。

2. 设计风险

一般而言，成熟技术的设计风险相对较小，但是如果设计工作没有遵照项目综合进度进行良好组织的话，潜在的设计变化以及图纸资料变更和延误依然是很大的挑战。对于首堆核电项目而言，设计阶段的研发和技术验证是风险控制的关键所在。对于新技术风险控制而言，关键是采取积极有效的措施来确保研发、设计改进和工程建设的连贯性。与此同时，新技术研发进度需要与工程建设进度相匹配。

3. 采购风险

制订采购计划应考虑制造周期和设备接口资料所需时间等因素。对于长周期设备而言，需要利用有经验的制造商。对于高风险设备，建议采取冗余采购方式（额外的设备可用于下一个项目或者增加的成本可接受）。来自新供应商以及采用新技术的设备是设计接口和设备采购管理延误的主要风险因素。执行过程的管理推演是进行风险识别的有效手段，包括技术、进度、成本和质量等方面的风险。然后可采取针对性的措施。在许多国家的核电开发过程中，出现了本地化需求。因此，应对设备本地化过程的风险进行充分评估。

4. 建设风险

大型核电站项目面临多种建设风险，如设计变化、图纸资料延误、资源匮乏（特别是合格工人）、事故、设备调试等，必须对这些风险进行准确评估，同时不应低估质量和安全要求对施工进度的影响。

第 12 章　在控制成本与复杂度的同时保证安全

建议

● 国际原子能机构（IAEA）制定的《安全基本准则》对于核工业的发展非常重要。根据这些准则，核安全的要求应考虑最新的科学技术进步。然而，在如何定义充分的核安全水平方面，IAEA 仍然非常笼统，"多安全才算安全？"仍然是一个尚待解决的问题。中法三院建议 IAEA 针对这一复杂问题给出更明确的意见。

● 很长时间以来，核安全都是在国家层面进行管制。然而，中法三院建议应在全球范围内统一核安全要求。正如 80 多年前航空业发生的情况那样，这种标准的统一是行业标准化的前提条件。在这一方向上首先需要对安全目标达成共识。

● 中法三院建议采用被诸多核能国家所使用的"风险指引"的方法，该方法可以平衡安全要求与核能益处二者之间的关系。中法三院对不管优劣只要求系统化应用"最佳可用技术"的方式表示质疑。

● 尽管中法三院认识到应定期对运行中的反应堆进行现代化改造，但建议在电站投入商业运行前的建设期间，安全要求应保持固定。必须全面地看待核安全，因此，即使采用了成熟的技术也不一定会在整体上产生积极的作用。

　　自核能开发伊始，安全就是重中之重。核安全的主要概念（包括运营者最终负责，独立的安全监管机构，设施审核和保护措施优化，通过纵深防御措施防止事故发生，通过多重独立屏障缓解事故，应急准备与响应）在核工业的早期就已确定。它们体现在 IAEA 发布的《安全基本准则》（2006 年最后修订）里。

　　因此，就核安全要求达成广泛一致是核电发展的前提条件。同时，这些核安全要求会随着运营反馈、科学技术进步以及对加强防护的社会需求提升而追加。因此，安全是一个发展的概念，是一个具有相对性的问题。系统性地追求更高的安全目标（无论它们意味着何种额外的复杂性）的做法是需要被质疑的。

核风险或核安全水平需要与其他社会风险相比较，并将其维持在一个合理的范围内。盲目地追求更高的核安全水平是不合适的。尽管在 US-NRC（美国核监管委员会）要求的推动下，全球在很大程度上就第二代反应堆的安全要求形成了共识，但是第三代反应堆重新带来了两大问题：①"多安全才算安全?"或者"是否有所有国家都认可的适当安全水平?"；②是否有可能制定一套各国都能接受的安全要求? 很容易理解，对问题①的肯定回答是回答问题②的前提条件。当比较不同国家近期核安全要求变化时，看起来各国的监管机构在这些问题上是存在分歧的。在承认绝对的安全并不存在以及核工业和其他行业一样将会面临"剩余风险"的同时，有必要通过明确的方法确定可接受的安全水平。

IAEA 安全目标和准则为此提供了答案，其优化准则是"针对有辐射风险的设施和活动采取的安全措施，如果在没有不恰当地限制使用，可在设施或活动整个寿命期内合理实现最高安全水平，则可被视为优化措施。"然而，如果假设风险为零，或者永远没有足够的安全性，那么优化意味着最大化，这可能会导致无休止的进程。

毫无疑问，核工业必须考虑最新的科学技术进步，但是也必须确保更好的设计优化，而不是一味地增加功能，提升复杂度。为了增加合理性，应采用"风险指引"的方法，该方法最初由 US-NRC 提出，并被 IEAE 正式采纳。遵照这种方法，可通过概率评估对新要求进行系统化分析，以确保能增强安全性。

针对超设计基准事故的最佳估算法是更现实的分析方法，并且避免了使用过于保守的方法。在对物理现象有充分理解的前提下，应接受更现实的设计方法，因为过分的保守设计不仅导致成本增加，而且将导致实施过程的延期和困难。

结论

不应仅仅通过采用风险分析来审视核能，而是应考虑同时采用风险和收益分析，然后在风险与收益之间进行权衡。

核安全是一个不断演进的过程，其进步是令人瞩目的。经历了多年的发展和经验积累后，核工业已经达到高安全水平，此外，核能也是低碳的。

需要实施基于物理理解和实验验证的合格仿真模型的方法，以避免采用过于保守的解决方案，从而提升核工业的竞争力。同时也需要与安全监管机构就技术事项进行互动，以使这些方法获得批准。

在全球范围内实现并深化安全规范的统一是一件恰逢其时的好事。

第 13 章　支持新兴国家项目筹备的国际路线的合理性

建议

为了能更好地支持新兴国家的核能规划,核电设备生产国应在以下两个方面取得进展:

- 取证及监管流程的标准化和稳定化;
- 相关政府保证实施长期电力采购合同的签订。

大多数新兴国家都急需稳定可靠的基础负荷电力供应,以开发基础设施,发展工业,满足大城市加速发展需求。核能可为满足这些要求提供潜在答案,它可与可再生能源互补,后者可在局部范围内满足部分用电需求。

目前,世界上有 50 多个国家正在积极研究开发核电,其中大多数为新兴国家,他们在核电建设方面缺乏经验,核电基础设施也欠发达。就此,国际原子能机构(IAEA)针对新兴国家核电基础设施建设颁布了一份指南(第 NG-G.3.1 号文件《国家核电基础结构发展中的里程碑》)。该指南指出,新兴国家在建设第一座核电站时通常会经历三个阶段,同时需要考虑 19 个具体的基础设施问题。结合国际经验,每个新兴国家可利用本指南选择适应其国情的核电发展模式。

此外,包括 IAEA 在内的国际核能界认为每个国家都能利用民用核能,但首先需要满足国际核能界已达到的核安全水平要求,为此应符合自身的时间表以及国情。所有核能国家在预防事故方面的相互依赖毫无疑问是一个制约。然而,当老牌核能国家帮助新兴核能国家克服困难开发核电项目时,这又变成了优势。中国和阿联酋的例子就是两种不同但又各自合理的核电发展路径。

中国由于国内市场规模可观，因此采取了引进、吸收、创新和大规模规划的核电发展模式。在 20 世纪 80 年代自主设计建设秦山核电站的同时，中国引进了法国的核岛技术和英国的常规岛技术，建成了大亚湾核电站。在此后的 20 年间，中国通过引进、吸收和创新逐渐形成了自己的技术体系，同时实现了标准化和量产。在此基础上，中国引入了美国、法国和俄罗斯的第三代核电技术（这些技术采用了先进的设计理念）。结合自身在核电站建设运营方面的成熟经验，中国形成了具有自主知识产权的第三代核电技术"华龙一号（HPR1000）"。在此过程中，中国在核电基础设施建设方面取得了长足进步，目前已具备核电全面出口能力。

阿联酋选择了全面引进的核电发展模式。阿联酋拥有充足的资本和具体的核电发展计划。为了快速发展核电，阿联酋根据其核能研究和相关技术资源匮乏的现实情况，选择了全面引入的发展模式。第一座核电站采用的是韩国技术。核电站的设计和施工由韩国负责，核燃料则采购自法国和俄罗斯。在人力资源方面，阿联酋的核能机构直接从海外聘请有经验的专业人才负责该国的核能监控。此外，阿联酋设立了由不同国家知名核能专家组成的国际顾问委员会，以推动核电基础设施的发展。

从普遍意义上讲，新兴核电国家面临以下三个不同层面的困难：

- 管理
- 工业
- 融资

第一类管理困难源于新兴国家人才核相关能力的欠缺，即便新兴国家只需扮演业主角色，这方面的需求也经常被低估。这种业主能力的建设要么主要以本国为基础（如中国），要么全面依靠外国人员和专家（如阿联酋）。

第二类管理困难是不熟悉核电项目的具体特点，但是这种困难可在 IAEA 或核电顾问公司的帮助下轻易克服。

第三类管理困难是由于核电项目是涉及整个国家的全面规划，有别于一般的工程项目。IAEA NG 指南里对此进行了全面说明。

然而，根据最新的项目反馈，在 19 个核电建设问题中有 3 个问题最为重要：

- 国家的安全和监管体系
- 稳健的融资机制
- 处置乏燃料和其他核废物的长期可靠的战略

这使得我们很容易理解对于新兴国家政府而言必须拥有核技术能力和长期承诺。

工业困难主要由于新兴国家的工业基础通常相对薄弱，或者不熟悉核电特点以及质量、可追溯性、审计和控制流程方面的高要求标准。另外，政府又通常希望尽快实现核电项目的本土化。

融资困难越来越频繁地发生在核电站的筹备工作进入创立业主公司这一关键阶段，可能出现在与股东、政府管理当局、融资手段、股东投资回报率上。这与下面一个或多个问题相关：

- 需要大量投资（数十亿欧元）
- 新兴国家融资能力有限
- 通过规范银行业的巴塞尔协议Ⅰ、Ⅱ和Ⅲ对潜在借贷人施加限制
- 新兴国家政府不愿保证长期为业主公司提供电价补贴

只有国际化路线也许能支持新兴国家决策者克服这些困难。

针对管理困难，双边政府协定和配套解决方案相对更适用于设立安全监管当局和监管框架、组织教育培训以建设班子能力以及推动长期技术支持能力的发展（如研究堆、热室和废物实验室等）。厂址选择研究也可被纳入这一范围，尽管核电设备生产商联盟也越来越多地负责进行场地勘察。

一期项目建设以及相关项目质量工作通常由政府部门与行业协会共同承担。最有效的途径是由预先选中的国有公司和有相应经验的外资公司共同承担，或至少与这些公司达成友好合作关系。

融资困难迄今为止主要通过核电设备生产国及其工业企业的深入合作来克服：

- 通过未来核电运营领导者（有时是反应堆生产商）的股权投资
- 通过贷款方或出口信用机构的借贷

BOO（建设-拥有-运行）模式和BOT（建设-拥有-转让）合同模式的采用（如土耳其项目中所用到的模式）是核电领域的真正创新，尽管目前这些模式主要被用于其他类型的发电计划。

相对来讲，长期电力采购合同对于确保投资回报而言是必不可少的。但是这种体制面临一些限制。即便大规模采用这种体制的俄罗斯现在也面临财政方面的制约。此外，在大多数国家，自由市场对电力价格的过度信心削弱了政府签订长期合同的能力，虽然也有英国上网电价补贴的例子。

全球取证流程的标准化和稳定化将会为投资者在核电项目规划中提供信心。

由政府担保并实施的长期电力采购合同的签订也会增强投资者对投资回报的信心。这些改进都依赖国际合作，但目前还没有建立相应的合作论坛机制。

最后，核电设备生产国开发的小型模块化反应堆在不久的将来可能是一种补充性解决方案，因为它们可大幅降低初始投资成本，简化并加速项目实施。这可能需要核电设备生产国与新兴国家之间加强合作（尤其是在法规、废物处置和保险方面）。

结论

就开发核电而言，不同国家应选择最适合其国情的发展模式，并通过国际合作逐步提升其核电开发能力。对于新兴国家发展核电而言，借助国际力量和进行双边合作是不可避免的。唯其如此才能克服相关的管理、工业和融资困难。

考虑到全球核工业的整体安全，拥有核工业供应链能力的国家（例如：中国和法国）应采取行动，通过分享其核能资源和帮助新兴国家安全有效地开发核电的方式服务全球社会。

第 14 章　核电活动 50 年来对人类健康
影响的评估

建议

● 商用核反应堆 50 年来的正常运行表明，其辐射影响是极低的，且远低于天然辐射水平。这一事实应更好地传达给公众。同等重要的是，应该报道煤等化石燃料的燃烧给健康所造成的影响，以提供正确的观点。

● 切尔诺贝利和福岛等严重事故所造成的致命辐射后果是有限的，但是，大片区域内不得不长时间疏散。因此，对第二代反应堆的改造是重要的，以进一步预防这类事故、缓解事故后果，除了在紧邻核电站周边地区有限时间的撤离外，无需采取任何应对措施。在这方面，应像中法两国所做的将第二代反应堆的安全水平改进到尽可能接近第三代反应堆的安全水平。

自 20 世纪 50 年代以来，核电生产对核电站运行人员和周围居民的健康影响，一直是许多研究的主题。这些研究全面评估核燃料循环正常运行时和发生事故时的辐射照射影响。联合国原子辐射影响科学委员会（UNSCEAR）的科学报告以及一些其他报告，定期总结这些研究结果。下面将讲述有关辐射照射的关注点。附录 D 简要介绍辐射照射的剂量水平及其健康效应。

14.1　正常运行时的辐射照射

每人每年所受到的天然辐射剂量为几毫西弗/年（mSv/a）左右。核电站辐射对附近居民的影响非常小，只有天然辐射剂量的千分之一到百分之一，并且它远低于将额外辐射剂量限制在 1 mSv/a 的公认 ICRP（国际放射防护委员会）标准。据 UN-SCEAR 报告，2000 年到 2002 年间，全球核燃料循环（从采铀到燃料制备到反应堆运行再到后处理）工作人员平均年辐射剂量约为 1.2 mSv，而 ICRP 规定的工人职业辐射限值为 20 mSv/a。

至于采铀行业，比如在世界第二大产铀国加拿大，铀矿附近没有发现超出正常本底水平氡的增加，采铀和选矿工人受到的平均辐射剂量约为 1 mSv/a，远低于加拿大法律规定的 50 mSv/a 的限值。

事实上，任何发电类型，在其全寿命周期的活动中都可能增加公众和工作人员的辐射照射，比如从采掘、建造、运行到后处理。重要的照射途径是来自土壤和地质结构中释放的天然放射性核素（氡及其子体）。例如：在获取建设发电站（不论其类型）所需的材料时，都需要开展采掘活动（尤其是需要采掘金属矿产）。UN-SCEAR 2016 年报告对不同类型的发电方式及其上下游活动的辐射照射水平进行了比较。

据估计，煤炭循环的辐射剂量占总集体剂量（个人剂量乘以受照人数）的一半以上，集体剂量是指一年中全球发电导致所有排放的辐射剂量之和，该评估建立在现代煤电厂排放假设的基础上。另一方面，包括发电在内的核燃料循环对集体剂量的贡献小于五分之一。正常运行情况下，就单位发电量产生的集体剂量而言，煤炭循环高于核循环，且除地热能外明显高于其他技术。

14.2　核电站及其他核设施事故伤亡及健康影响评估

从开始用核能发电以来，全世界发生过几次重大事故（包括核电站堆芯熔毁），伴有大量释放的后果。

三里岛事故发生于 1979 年 3 月，当时引起了堆芯部分熔毁，导致放射性气体（0.55 PBq 的放射性碘）释放到环境中。根据美国核协会使用的官方放射性释放数据，"居住在核电站方圆十英里内的居民受到的平均辐射剂量为 0.08 mSv，任何一个人受到的辐射剂量不超过 1 mSv。"各种流行病学研究结论表明，该事故没有造成明显的远期健康影响。

切尔诺贝利事故发生于 1986 年 4 月，当时引起了全部堆芯熔毁，导致放射性气体（1760 PBq 的放射性碘，85 PBq 的放射性铯）和物质释放到环境中。

134 名应急工作人员罹患急性放射病，其中 28 人死于辐射。在受到中等剂量照射的恢复清理工作人员中，发现白血病和白内障危险增加的一些证据。在儿童和青年期受到照射的人群中，甲状腺癌的发生显著增加，其归因于在事故早期饮用了放射性碘污染的牛奶。甲状腺癌是一种少见的疾病，且其预后已得到极大改善。1991 年至 2005 年间，切尔诺贝利周边污染地区报告病例数超过 6000 例，其中 15 例被证明是致命的。

除了儿童期照射导致的甲状腺癌外，未发现被污染地区的居民过多地患上任何其他实体癌和白血病。UNSCEAR 委员会曾指出"大多数地区的居民受到与年平均天然本底辐射水平相差无几或只比其高几倍的低水平辐射照射""不太可能在普通人群中导致严重的健康影响"，但是"该事故导致的严重破坏会产生重大的社会影响和经济影响，给受影响的人群带来巨大的痛苦"。

福岛核事故发生于 2011 年 3 月，由地震引发的海啸导致，引起福岛沿海地区 6 个核电机组中的 3 个堆芯熔毁，放射性气体（少于 500 PBq 的放射性碘，少于 20 PBq 的放射性铯）和物质释放到环境中。

未发现由辐射直接导致的急性健康影响（包括死亡）。在事故发生后的第一年内，参与事故善后的工作人员和生活在周边的成年人接受的平均剂量分别约为 12 mSv 和 1~10 mSv，婴儿约翻倍。据估计，在头 10 年内产生的剂量为第一年产生剂量的两倍。UNSCEAR 报告（2013 年）和白皮书（2015 年、2016 年）对事故发生以来的相关研究进行了评述。委员会认为"在辐射照射剂量最高的儿童组中，理论上，出现甲状腺癌危险增加的可能性是存在的，然而，甲状腺癌在儿童中是一种罕见疾病，从统计上看，预计在这一人群中不会发现可观察到的影响。"

UNSCEAR 委员会也曾指出"在普通公众和工作人员中观察到的最重要的健康影响，体现在心理健康和社会福祉上。"

该表述在每次发生放射性物质大量释放时都是合适的。因此，重点是改造第二代反应堆，从而不仅更进一步降低堆芯熔毁的可能性，而且允许在安全壳超压时，能够通过安全壳过滤排放系统和其他配置，使裂变产物的释放受到控制。改造后，这些反应堆的安全水平将与目前第三代反应堆的安全水平尽可能接近。如果日本曾考虑采用这种系统，福岛核电站内及周边就无需长期疏散。中国和法国的核电站就装有这种系统。

结论

在过去 50 多年的核电站发展历程中，从严重事故的经验反馈中可以看出，如果裂变产物是被封闭的，那么其对人类无任何辐射影响（即使有，其影响也非常有限）。但另一方面，如果有裂变产物泄漏，就必须进行大范围疏散。核电站事故对公众的健康影响亦非常有限：电离辐射对健康的影响取决于所受到的辐射剂量，如前面所述，这种剂量是非常低的。但在事故地点附近，心理健康和社会福祉方面的社会影响却很严重。因此，需要付出大量的努力提高所有在役和未来反应堆的安全性，以在任何情况下阻止放射性物质的释放。核电站正常运行情况下，对公众的辐射照射水平是非常低的，甚至低于燃煤发电等活动产生的辐射水平（煤中含有天然形成的微量放射性元素，如铀、钍、氡等，在火电厂燃煤发电时会逃逸到大气中）。

附录 D　辐射照射剂量水平和健康效应

无论过去还是现在，人们每天受到非电离辐射的照射，这种辐射的能量不足以使原子或分子电离，这种非电离辐射的例子有微波炉、收音机或手机附近的电磁场。有些形式的辐射——通常被称为电离辐射，有足够的能量使原子的电子被击出，破坏了物质的电子/质子平衡，形成正离子。

人类不断受到始终存在于环境中的天然辐射源以及辐射应用中的人工辐射源的照射。全球人均年剂量估计约为 3 mSv（表 2），其中约 20％的照射来自人工源，主要来自医学应用，例如，一次腹部计算机断层扫描（即 CT）的剂量约为 10 mSv。

表 2　各辐射源对公众照射的全球人均年平均剂量（《辐射效应和源》，UNEP，2016 年）

总的天然源	2.42 mSv	总的人工源	0.65 mSv
食物	0.29 mSv	核电站	0.0002 mSv
宇宙射线	0.39 mSv	切尔诺贝利事故	0.002 mSv
土壤	0.48 mSv	核武器试验落下灰	0.005 mSv
氡	1.26 mSv	核医学和放射诊断	0.65 mSv

电离辐射的能量可通过杀死细胞或诱变细胞导致活体组织的损害。这些健康效应依赖于辐射照射的剂量。在 UNSCEAR 报告中，将超过 1000 mSv 的剂量称为高剂量，100～1000 mSv 的称为中等剂量，低于 100 mSv 的称为低剂量。

如果辐射照射杀死的细胞数量足够多，可导致组织反应甚至死亡（如脱发、皮肤烧伤和急性放射病等）。当剂量超过一定阈值时，这些效应的严重程度会随着剂量的增加而增大。例如，大于 1000 mSv 的剂量可导致急性放射病。可能有害的人体组织反应的剂量阈值高于 100 mSv。这种效应见于辐射事故或放疗中。

如果辐射引起的细胞变异不能修复，则可能导致癌症或子代的遗传疾病，这些效应是随机的，其发生概率依赖于所受到的辐射剂量。UNSCEAR 2010 年报告指出"充分的流行病学证据表明，人体受到中等和高剂量辐射照射，可导致很多器官实体肿瘤和白血病的超额发生。"例如，截至 2000 年 12 月，广岛长崎原爆幸存者中 10423 人死于癌症（实体肿瘤与白血病之和），其中 572 例（约 5%）可归因于爆炸导致的辐射照射。然而，没有明确的人类证据表明辐射照射的超额遗传效应。

第 15 章　相对于实际危险的风险认知

建议

● 应该加大力度推进和落实核安全文化，加强双向沟通，帮助公众以理性和客观的方式看待核能风险。

● 应该以通俗的语言说明福岛核事故后，为加强现有第二代核电站安全性、改进第三代和三代＋核电站设计所采取的措施。对严重核事故应特别强调：首先，在采取相应预防措施后，该类事故的发生概率降低了一个数量级；其次，该类事故仅在短期内会对厂址附近产生一定的辐射后果。

每当新技术出现时，都会出现两种相对立的思维方式。一种思维方式可简化为对成本/收益平衡的评估（即运营者和工业参与者的思维方式），他们必须不断创新以保持竞争力。另一种思维方式注重的是技术创新的负面影响，并试图重新构建一种所谓"理性"的方法，即通过考虑更多伦理、定性或间接因素，对成本/收益评估结果强加限制。

这种思维方式要求工程师和科学家们要不惜一切代价避免灾难的发生以及灾难导致的负面影响。世界上大多数国家的证据表明：核活动正以此为目标，同时也是对上述观点的明显例证。

因此，有必要强调，虽然风险依赖于危险而存在，但不应该将风险与危险相混淆。风险是危险与其暴露量的乘积。如果没有暴露于危险之中，就不存在风险。

可以想象，一旦风险被客观证实，风险认知就会与风险本身的客观评价相符合。但这是理想情况，可以从以下三个角度来分析。

（1）心理学。风险认知不只是依赖于非理性因素。每个人都建立有自己的"风险篮子"，并以特定的理性来管理它。例如，某些风险会因其提供的补偿而被接受，核电站或核设施周边的居民就是这类情况的实例。

（2）社会学。风险认知在很大程度上取决于文化、国家和历史因素。此外，个体选择的风险与集体经历的风险在引起恐惧的方式上是不同的。核能并不是由其弊端可能影响的大部分群体所明确选择的一种发电方式。

（3）传播学。风险传播的主体会影响公众的风险认知与风险价值判断。由于核能具有较高的技术性和科学性，大多数公众对核能缺乏认知经验；可以预见，在当今网络媒体时代，媒体的影响无处不在，其对公众核能认知的形成具有启发式作用。

15.1 从心理学的角度分析核能风险认知

心理学认为个体通过心理认知机制感知风险，风险认知偏差是个体在认识和判断风险时所发生的某种偏离。导致风险认知偏差的原因包括：个体的人格特征、知识经验、风险"损失"或"获益"预期等主观因素，以及风险的性质、大小、可控程度和易了解性等客观因素。因此，不同的人对核能风险的认知是存在差异的。对于专业人员，通常基于技术评价来权衡利弊、判断风险；另外，公众往往倾向于感知与其切身安全、健康利益紧密关联的风险，而忽视核能的效益，这也是核设施选址中的邻避效应的主要原因之一。

公众往往将潜在灾难、不可控、未知等作为核能的直观特征，而三里岛、切尔诺贝利、福岛核事故加剧了公众对核能的这种心理认知。虽然核设施正常运行的辐射影响远低于煤炭行业产生的辐射影响，并且第三代和第四代反应堆在设计上考虑了防止任何情况下放射性物质的释放。然而，公众仍然认为核能风险远高于其他行业。

相关政府部门、研究人员、媒体对中国公众核能接受度的调查结果显示：

● 公众对核能信息的了解渠道有限，缺乏基本的知识；

● 由于公众参与度和信息透明度不足以及对核安全的担忧，大部分公众对发展核能普遍存在疑虑，在中国只有约40%的公众支持发展核电；

● 在福岛核事故后，公众对核电项目更为敏感（尤其是对家门口的核电项目反对比较强烈）。

在法国，《IRSN 2016 年法国公众对风险和安全认知晴雨表》的主要调查结论是：

● 恐怖主义已经成为第一担忧；

● 超过半数的法国人认为，相比于十年前对科学更有信心，并且他们相信核领域的专家、技术和科学组织；

● 大多数人主张让公众接触专业的结果；

● 尽管 46％的调查对象认为所采取的预防措施确保了法国核电站的高度安全，但约 90％的人认为核电站事故会产生非常严重的后果。

15.2 从社会学的角度分析核能风险认知

个人对风险的认知不仅仅是基于个体的心理认知，而且是与社会组织或社会制度相联系的，对风险的判断受其所处社会地位、群体和背景的影响。灾难事件与心理、社会、制度和文化状态等相互作用，会影响个体对风险的感知，其表现为风险认知上的社会连锁效应，即当人们对恐惧的描述在交谈中被引用得越频繁，公众越关注与此相关的信息，使得人们"感知"（想象）的风险与实际产生偏差。例如：在日本福岛核事故后，美国、法国、德国等国家出现了公众抢购碘片的现象，马来西亚、菲律宾和俄罗斯公众抢购碘酒，韩国公众抢购海藻产品，中国公众抢购碘盐。更重要的是，有核能项目因公众抗议而被叫停的情况。针对这种情况，需要仔细研究以更好地应对此类社会危机。

15.3 从传播学的角度分析核能风险认知

风险传播的主体包括国家政府机构、专家学者、社会公众、媒体、非政府组织等。不同传播主体对风险传播效果的影响不同（尤其是对公众风险认知与风险价值的判断）。切尔诺贝利事故后，政府、媒体和不同组织对事故人员伤亡情况的报道相差甚远，导致公众对事故的实际影响无法做出正确的判断。

公众对于风险信息的获取与信任是相当敏感的。风险信息不完整往往会加剧公众对风险的疑虑，甚至丧失对信息的信任，例如：政府对于风险处理的回避性措施或模棱两可的态度。媒体将福岛核事故的"氢气爆炸"宣传为"核爆炸"，也在一定程度上加剧了公众的恐慌心理。

中国致力于完善国家核安全体系、提高核安全能力和培育核安全文化。2014 年中国正式将核安全纳入总体国家安全体系。

15.4 核能风险认知存在的问题、挑战和困难

风险既客观存在，又受媒体传播、个体认知和判断、社会、制度、文化状态等不同层次的影响。随着世界范围内的核电建设与核事故的发生，及各国核电政策的变化与发展，公众对核电的认知也随之变化，呈现出关注度高、认知与接受度有限、态度主观与非理性的特点。如何提高公众认知水平、规范信息传播和降低公众恐慌情绪是世界各国核能发展所面临的挑战和困难。

为了更好地理解如何改善公众意识，并鉴于上述考虑，有必要强调核活动中风险认知与实际危险之间的两个主要差异。

● 实际危险：如果三代＋核电站发生严重事故，不会产生重大的辐射后果（并且只需要采取有限的应对措施）。

然而，公众对该风险的主观认知是：在发生飞机撞击或损坏反应堆堆芯的严重事故情况下，仍会有大量的放射性物质从安全壳中释放。

● 实际危险：低放射性水平和低剂量的有害性仍然是生物学研究的一个重要课题，目前尚未得到明确的结论。但是，生活在高本底辐射区域（印度喀拉拉邦、法国布里塔尼地区等）人群的经验表明，未显示出可感知的长期影响。

然而，公众对该风险的主观认知是：来自核电站或核设施的照射对人类健康是危险的。但是，对于牙科或医院的 X 射线检查的照射，却不会对人类健康产生危害。

结论

　　关于核科学技术研究成果完全透明和公开的政策取得了一定成效，其极大地有助于缩小实际危险与公众风险认知之间的差距。然而，由于公众的核科学技术知识有限，存在普遍高估核能实际危害的情况。仍需要开展相关工作，使公众意识到在切尔诺贝利和福岛核事故后，全球核反应堆采取的措施在安全方面取得的巨大进步：通过采取充分的预防措施，法中两国现有反应堆以及第三代反应堆的堆芯熔融概率已降低十倍。此外，现有反应堆改造或新反应堆设计中采取的缓解措施大大降低了此类事故的辐射后果。因此，除短期内在核电站周围需采取措施外，无需采取任何其他应对措施。

第 16 章　公众认识和管理要求提升

建议

● 发展核电的国家必须共享信息和观点，以解释核能的优势和弱点，并采取能够提升公众意识的补救措施。必须强调核能是稳定可靠的基荷来源且不会产生温室气体排放。

● 核电运营者和利益相关方应就核电站的成功运行采取积极的沟通策略，在通报事件时透明公开。

● 除共享信息和观点之外，沟通策略还应考虑大规模开发间歇性可再生能源的国家的不同发展速度。如果可再生能源的利用是为了替代使用化石燃料的电厂，而不是为了减少已经是零碳排放的核电，那么可再生能源大量开发与核电并不冲突。

无论是工业化国家还是新兴国家，西方国家还是亚洲国家，综合全球情况来看，核能的优势都包括：

● 不排放温室气体，有助于应对气候变化；

● 可提供大量基荷，确保电网稳定，同时允许在国家层面发展工业和大型基础设施；

● 相比其他发电能源，核能如今储量丰富，能稳定供应低价电能，保证能源安全；

● 无需储存电能，与间歇性能源相比，核能具有绝对优势。

但核能也有它的弱点：

● 高放射性长寿命废物负担切实存在，其数量相对有限，但能留存数十万年；

● 严重事故对大面积区域和大量人口的影响。

在所有这些国家，提升公众意识的基本工作应是，就以上每个条目为公众提供容易获取的信息，并要特别注意量化问题和讨论解决方案。

每个国家的国情都有其独特之处，不可一概而论，电能增长的趋势就是一例。对中国和其他快速发展的经济体而言，因为用电需求快速增长，它们可以大规模、渐进式地发展间歇性可再生能源发电能力，但不会阻碍核能部门的同步发展。欧洲的情况就很不一样，这里的用电需求相对平稳（然而，如果化石燃料消费被电能取代，电能需求可能会再次上扬）。因此，快速、大规模地发展间歇性无碳可再生能源（太阳能和风能）虽然需要大量投资，但主要后果是拉低核能发电的需求。然而，这对二氧化碳排放的影响非常有限，因为核能同样不排放二氧化碳。不仅如此，可再生能源对核电装机容量的影响也非常有限，核电装机容量必须保持在一定水平，从而为间歇性可再生能源提供必要的备用电能。

无论主张发展间歇性能源"可供迈向未来的世界"的流行理念，还是类似"负增长理论"的所谓哲学思想，都使得欧洲的困境雪上加霜。欧洲需要新的论据和进一步的努力去提高公众理性化的认知。

与前面提到的 5 个条目对应的进一步论据和详细内容，可在本报告的其他章节中找到。特别提到的是，与放射性废物管理有关的问题在第 6 章中讲述，该章节得出的结论是，长寿命放射性废物深地质处置的方法可行。第 14 章回顾了与核电站和其他设施有关的事故，同时还简要介绍了当前的核电站已经实现的巨大改进，以及第三代系统在设计方面做出的改进。当然，为促进公众对欧洲现状的了解，更多思考仍是有益的。

目前以及可能在未来数十年中，如果没有芬兰和英国等国政府的强有力支持，那么气候变化问题恐怕不足以说服公众和政界转变对新建核电站的否定态度。这种政府支持不仅在中短期内的决策阶段（5～10 年）和建设阶段（5～10 年）是必须的，而且在长期（例如：核电站寿命期（至少 60 年））也是必须的。

如果说公众知道核能对减少二氧化碳排放的贡献（但许多反核 NGO 总在或多或少地否认这一贡献！），那么其价值恐怕并未得到公正的评价。由于更严格的安全要求已导致核能发电成本上升，而光伏发电成本相比之下却已大幅下降，因此核能在发电成本方面的经济效益（同样，至少欧洲是这样）一直备受质疑。

如今，政治家、媒体和公众的技术、工业和经济文化往往都相对有限，且以短期为导向。在与一些重要决策过程中的政治甚至是哲学观点相权衡时，工业资产和经济因素通常都会被低估。

所幸还有一些政治决策者能考虑到经济目标和技术限制。隶属于法国议会科技选择评估局（OPECST）的许多法国议员就是这样的决策者。该评估局致力于开展全面透彻的调查，并发布高质量的报告，此类报告着重提供通常被公众和媒体忽略但在决定能源领域的复杂问题时应该考虑到的信息。

然而，几乎所有这些国家对核能的主要误解都在"能源转型"的步调上。如果间歇性可再生能源需要通过化石燃料发电来提供备用电能的话，那么我们没有理由要求在50～100年内用间歇性可再生能源替代核电（如果间歇性可再生能源需要化石能源作为备用）！这是必须传递给公众的附加核心信息。

核电运营者和反核NGO的沟通策略对公众认知和管理要求都有很大影响。应格外注意避免核电运营者的信息被解读为完全违背原意的情况。例如，EDF的宣传语"安全第一"被许多人解读为"EDF核电机组目前不安全"，需要紧急整顿。这种解读尤其被人们与所谓的"大整修"项目需要的巨额投资联系到一起。

要想新开辟新厂址建造核设施，一方面需要强有力的政府支持，另一方面，制定大量协调一致的程序也特别重要。这主要不是因为当地反对者，而是专业的国际反对者，以及至少在法国是如此——错综复杂的环境法律网络，例如：水法、海岸线法和生物多样性保护法等。甚至现有厂址扩建也可能出现问题。在中国，内陆项目政策在福岛核事故发生后仍然未放开。

结论

核能的益处，尤其是它能稳定、大规模地生产电能，同时温室气体排放量低，未得到充分宣传。公众不够了解基荷发电的需要，对于电网提供的卓越服务，它的稳定性，以及它对气象条件的较低敏感性，公众同样缺乏认识。有必要在讨论中加入更多有关能源结构的现实问题。特别应强调，间歇性可再生能源发电需要补偿：当可再生能源不能制造电能时——例如没有阳光或风时，备用容量必须立即上线，因为目前尚无大规模电能储存设施，而且这种情况可能会持续多年。如果传统备用容量是依赖化石燃料燃烧，那么它会带来温室气体排放。此外，当可再生能源发电量超出临界水平时，大片相互连通的电网将会出现严重的稳定性问题。要求传统发电技术以高度间歇性的方式运行，在技术或经济上是不可持续的。

有必要向公众解释这些基本问题，并说明核能可以安全持续地为大型、互联的电网供应电能，从而满足城市人口不断上升的用电需求。因为有关能源结构的主张通常易受情感因素的影响，因此需要加强公众对于技术限制和基本经济目标的了解，同时要进一步提供针对安全问题以及在安全处理事故方面已取得进展的信息。要想做到这一点，一方面需要制定更好的沟通策略，另一方面需要加强能源问题的教育，从而使得人们能够了解问题的真相，而不是盲目听信误导性言论，也使公众能依据理性的事实分析形成自己的观点。

此外，还必须考虑国与国发展速度不同的问题，例如：如果一个国家经济发展缓慢，就可能会对大规模发展间歇性可再生能源与核能构成冲突，而在经济发展迅速的国家，这种冲突就不一定成立。

第 17 章　增进公众理解的组织、方法和不同利益相关方角色

建议

要想提高公众对核能的接受度，建议从以下三个不同层面采取行动：

● 在技术层面，必须充分考虑从以往重大核事故中吸取的经验教训，以确保安全高效地运营现有在役核设施。

● 在组织层面，中央政府应制定明确的能源战略，并坚定实施。同时，必须将核设施运营者与安全监管部门及其支持机构明确分隔开。在涉核项目所在地建立一个公开、平等、参与便捷的沟通对话平台，这将有利于提高公众信心。

● 在沟通层面，应当尽最大努力让广大公众能够了解到公开透明、准确以及立体结构化的信息。要想提高社会接受度，重点要加强有关能源问题、基本科学知识、经济因素、环境影响和风险方面的教育，并提高公众对重大能源挑战的认识。

从三里岛、切尔诺贝利和福岛这几起重大事故中吸取的经验教训，让我们重新思考，在不得不面对放射性物质大规模环境释放时，要想做好更充分的准备所必须考虑的风险因素，同时也凸显出在核问题方面向公众提供独立、透明化信息的必要性。而这一切需要在不同层面采取行动。

第一个层面的行动基本是技术上的，它包括：不断改进电厂运行的效率和安全性，在当前的核电站机组中和未来电厂的工程设计中，充分吸取过去的经验教训。因此，必须对风险因素进行合理的分析和排序，以便更有效地管理它们。这已催生出第三代反应堆，它在抵抗事故和各种攻击行为的能力上有了大幅改善。

第二个层面的行动是组织上的，它涉及三个级别的利益相关方：

- 中央政府及其能源、教育、宣传、卫生等职能部门；

- 核工业的所有利益相关方，包括监管部门、经营者、安全机构、独立的技术支持机构、行业协会等；

- 涉核项目所在地的地方政府以及社会公众。

对于不同的利益相关方，可以赋予不同的职责和可采取的行动：

（1）中央政府应当科学统筹，制定发布本国的能源发展战略（包括发展核能），并协调整合下属的各职能部门全力服务于国家能源战略的实现。

（2）对于核行业的各利益相关方，必须明确界定其各自的角色和职责，以确保他们能够践行健康的核安全文化。IAEA 指南对此的表述是："安全文化是存在于组织和个人中的各种特性和态度的总和。它建立一种超出一切之上的观念，即：核电站安全问题由于其重要性要保证获得应有的重视。"要想建立这种安全文化，就必须明确各相关方的职责：

- 核反应堆的运营者担负安全的主要责任。

- 一个独立公共安全机构负责监管所有民用核设施的运营。在法国，它的独立性要通过对其成员进行六年不可撤销的委任提名来保证。该机构负责组织对核电站和设备进行定期、强制性的检查，发放运营许可证，并有权在其认为有必要的情况下中断核设施的运行。

- 行政监管部门需要依靠强大的专业技术。它可以在内部进行培养，或者由一个或多个技术支持机构提供支持（参见第 7 章）。

（3）在地方层面，国家的权力下放有着不同的情况。比如，在法国，中央政府在地方上的代表（区域行政长官）是经营者的唯一对话者。地方政府机构专门负责沟通协商及告知涉核项目所在地区的民众：与核能有关的项目都是按照完整的决策程序规划的。项目获批后，地方政府机构应当与经营者一起积极建立针对民众的便捷的沟通对话平台以及透明的信息披露机制。此外，地方政府机构应积极探索实现核项目与地方经济和社会协调发展的方式，包括通过核项目为财政做贡献、提升当地的就业率等。

在有些情况下，做决策之前必须组织开展公众意见征询，以便更好地了解恐惧的来源。法国的一个被称为 CIGEO 的核废物深地质处置规划项目就属于这种情况。

第三个层面的行动是，通过提高公众意识与认识，让公众知道已经做出的这些努力。

然而，因为缺乏基本的辐射知识教育，加之辐射是看不见的东西，并且许多人会联想到核武器，对核能的恐惧就会加剧。即使核物理有着最显而易见的益处（例如：肿瘤放疗，基于核磁共振的医学影像，通过放射性同位素了解地球历史），但人们还是以太恐怖为由反对提到带"核"的词汇。已经很清楚的是，核行业的未来发展在很大程度上取决于公众对这项技术的认识和接受度。欧洲在这方面就有活生生的例子，比如，意大利目前拒绝接受任何核工业，还有德国，在福岛核事故后已经决定，在2022 年之前关闭其核电站。因此，关键要解释清楚在新一代的核电设计中已包含了这些安全措施，而这需要确保信息透明，开展必要的基础教育，并尽可能响应广大公众对信息与日俱增的需求。此外，还需要改进我们在事故或事件（无论是自然的还是人为的）中及时告知公众，以及与公众沟通的方式。在法国国家层面，一个核安全信息透明高层次委员会（HCTISN）将保证为公众提供有关民用核技术的准确和可获取的信息。

结 论

除了确保核电站运行安全所采取的大量行动外，明确不同利益相关方的职责并提供透明、准确和结构立体化的信息是对发展核能不同声音的最佳回应。此外，还要强调核能的益处，它是一种安全、清洁、高效的电力能源，同时也是推动经济发展的工具之一。

词 汇 表

ABWR：advanced boiling water reactor，先进沸水反应堆

APWR：advanced pressurised water reactor，先进压水反应堆

ASN：Autorité de Sûreté Nucléaire（France），法国核安全局

BOO：build own operate，建设-拥有-运行

BOT：build own transfer，建设-拥有-转让

BWR：boiling water reactor，沸水反应堆

CAD：computed aided design，计算机辅助设计

CEA：Commissariat à l'Energie Atomique（French Atomic Energy Commission），法国原子能委员会

CEFR：China experimental fast reactor，中国实验快堆

CGN：China General Nuclear Power Corporation，中国广核电力股份有限公司

CNNC：China National Nuclear Corporation，中国核工业集团公司

COP：Conference of Parties to the United Nations Framework on Climate Change，《联合国气候变化框架公约》缔约方会议，联合国气候变化大会

DOE：Department of Energy，US，美国能源部

EDF：Electricité de France（French Utility），法国电力公司

EPR：European pressurised water reactor，欧洲压水式反应堆

ESBWR：essentially simplified boiling water reactor，经济简化沸水反应堆

EUR：european utilities requirements，欧洲用户要求文件

FNR：fast neutron reactors，快中子反应堆

FOAK：first of a kind，同类别首个

GCR：gas cooled reactor，气冷反应堆

Gen-Ⅱ，Gen-Ⅲ，Gen-Ⅳ：二代、三代、四代，指目前正在运行的或正在部署

的第二代、第三代和第四代核反应堆。第一代为原型，现已退役

GFR：gas cooled fast reactor，气冷快堆

GHG：greenhouse gas，温室气体

GIF：Gen-IV International Forum，第四代核能系统国际论坛

HL-LLW：high level-long lived waste，高放长寿命废物

HLW：high level waste，高放废物

HPR1000：advanced pressurized water reactor developed in China (also named Hualong One)，华龙一号，中国开发的先进压水反应堆

IAEA：International Atomic Energy Agency，国际原子能机构

ICRP：International Commission on Radio Protection，国际放射防护委员会

INSAG：International Nuclear Safety Advisory Group (advising the IAEA general director)，国际核安全咨询组（为国际原子能机构总干事提供咨询建议）

IRSN：Institut de Radioprotection et Sûreté nucléaire (France)，法国辐射防护和核安全研究院

IT：information technology，信息技术

LLW：long-lived waste，长寿命废物

LLW：low level waste，低放废物

LWR：light water reactor，轻水反应堆

MOX：mixed uranium-plutonium oxide，铀钚混合氧化物

MSR：molten salt reactor，熔盐反应堆

NEPIO：Nuclear Energy Programme Implementing Organisation，核能项目实施组织

NGO：Non-Governmental Organisation，非政府组织

NNSA：National Nuclear Safety Administration (China)，中国国家核安全局

NPP：nuclear power plant，核电站

NRC：Nuclear Regulatory Commission (USA)，美国核能管理委员会

PLM：product lifecycle management，产品生命周期管理

PPA：power purchase agreement，购电协议

PWR：pressurised water reactor，压水反应堆

R&D：research and development，研究与开发

RW：radioactive waste，放射性废物

SBO：station blackout，全厂断电

SFR：sodium-cooled fast reactor，钠冷快堆

SG：steam generator，蒸气发生器

SMR：small modular reactor，小型模块化反应堆

SWU：separation working unit，分离功单位

TMI：Three Miles Island，三里岛

TNR：thermal neutron reactor，热中子堆，热堆

TSO：technical support organisation，技术支持机构

UNSCEAR：United Nations Scientific Committee on the Effects of Atomic Radiation，联合国原子辐射效应科学委员会

URD：utility requirement document（developed in the United States），美国用户要求文件

VHTR：very high temperature reactor，超高温气冷反应堆

后　记

在当前大部分电力依靠化石燃料（尤其是依靠煤炭生产）的大背景下，核电是以安全、高效、清洁的方式供应能源，同时又是解决环境和气候变化问题的最现实选择之一。由于它是一种稳定、丰富的能源，因此能可靠地供应可调度电力，并对波动性大、不易于调配以满足电力需求的可再生发电能源（如风能或太阳能）形成很好的补充。

发展核能依然面临着诸多与安全性、长寿命放射性废物管理、先进核反应堆的开发与部署、经济性、公众的接受程度等有关的挑战和问题。作为具有世界前列的核电站装机容量的国家，中国和法国都非常重视全球核能的和平利用，并且有责任且有意愿帮助新兴国家建设核电站，共同应对他们将面临的挑战。

秉承第 21 届和第 22 届联合国气候变化大会（COP21，COP22）致力于显著减少全球温室气体排放量的精神，中国工程院、法国国家技术院和法国科学院相信，他们共同提出涉及某些核电复杂问题的倡议，能向其他国家的学院、决策者乃至社会各界提供有价值的信息。

本报告反映了中法三院作为独立机构的立场，不应被解读为核电从业者或中法两国政府的立场。在本报告中，中法三院旨在勾勒出核能发展的历史与前景，辨析需要考虑的关键问题，以让核能变得更加安全经济，惠及发达国家和新兴国家。这篇报告虽然涉及了诸多问题，但并非详尽无遗。它汇集了过去六个月我们反思的心得和讨论的成果。

本报告包含一篇综述和十七章。简介后的两个章节简要介绍核开发的历史、问题与挑战，并讨论与部署第三代核电站有关的问题。接下来的两章论及科学层面，讨论未来反应堆设计的前景与挑战（包括第四代反应堆的现状以及小型模块化反应堆理念）；接下来的一章介绍了核废物的管理。然后讨论了技术问题以及安全问题，指出了对安全技术支持机构的需要、数字化以及新型设计工具的进步与挑战；强调了核研究、核设施与核基础设施的重要性；审视了人员的教育和培训问题，其目的一方面是吸引年轻的高校毕业生进入核工业，另一方面是培训从业人员，使他们具备足够的科学背景以及安全管理文化。有两个章节讨论了工程问题，包括管理核电项目、在控制成本与工程复杂度的同时满足安全要求的问题。此外，本报告还审视了为新兴国家核项目提供国际支持的合理性。

最后四章则讨论了社会问题。其中一章提供了对过去 50 年来全球核电活动对人体健康的影响进行的评估。由于安全是核电站运行的核心问题，因此有必要分析风险认知并与实际危险进行比较。本报告强调了提升公共认知的必要性，并且考虑了所需的管理要求以及在增进公众理解和控制法规不断增加的风险（因为法规的增加会使核设施的发展和运行变得复杂化）方面的组织和不同利益相关方所扮演的角色。

Synthesis and recommendations

France and China are countries with large and lasting nuclear power plant capacities. France has a long experience in operating nuclear reactors and all the facilities of a closed fuel cycle. China is the country where the increase in nuclear power is foreseen to be the largest in the world for the coming years. Both countries are engaged in an electrical transition phase to optimise their "energy mix" and both have research and development (R&D) programmes to prepare the next generation of nuclear systems.

In line with their interest in sustainable energies and acting as independent bodies, the three Academies (Chinese Academy of Engineering, National Academy of Technologies of France and French Academy of Sciences) have analysed some issues and challenges raised by nuclear energy. They came to the following considerations and recommendations.

1. Decarbonising the energy systemraises difficult issues that should not be underestimated. It is generally considered that this will require an increase in the contribution of the electric vector (terrestrial mobility, industry, urban uses, etc.). Nuclear energy constitutes one of the most realistic options for supplying electrical energy. The latest generation of nuclear power plants (NPPs) constitutes a highly realistic option for supplying electrical energy in a safe, efficient and clean way and simultaneously solving environmental and climate change problems. They can reliably provide dispatchable electricity and complement renewable energy sources (like wind or solar photovoltaic) that are mostly intermittent with a fluctuating production independently of the demand and are not easy to dispatch to respond to demand.

2. As nuclear energy is highly concentrated in large NPPs, it has a smaller footprint (requires a smaller amount of land) in comparison with more dilute energy

sources. NPPs are capable of delivering the energy that is needed by the present megacities and by those that will arise in the future. Nuclear energy uses uranium, which is an abundant resource, without any cartel manipulating its availability on the market, thus ensuring security of supply.

(1) In the foreseeable future, there will be no solutions able to economically store large quantities of electricity. Other means of compensating for intermittent productions are therefore necessary to ensure that even more fluctuating demand can be met with large response reactivity and deployment, and thus satisfied at all times; however, the goal of decarbonisation would be missed if fossil fuels were to provide the necessary back-up.

(2) There are technical and economic limits to the share of intermittent energy in large electric networks (network-stability, recovery plans in case of large scale black-outs, etc.).

3. Nuclear power is still a relatively young technology in continuous progress.

(1) Major accidents have been analysed; return of experience and lessons learned have led to significant changes in the design and operation of reactors.

(2) The latest generation of reactors (Gen-Ⅲ) that is now under construction has been designed to make sure that there will be essentially no significant radiological consequences beyond the boundary of the nuclear site, even in case of a severe accident with fusion of the reactor core.

4. Other paths of progress are anticipated and their development should be encouraged.

(1) Gen-Ⅳ reactors, including sodium-cooled fast reactors (SFRs), which have the capacity to make full use of uranium and allow multi-recycling of spent fuel and transmutation of very long-lived radioactive elements present in radioactive waste.

(2) Small modular reactors (SMRs) that are adapted to small electric networks. Their modular construction allows to alleviate the complexity of large construction sites.

5. Accordingly, it is important to maintain ongoing R&D efforts targeted at reducing cost and risk, which should include:

(1) Sharing of human resources and research infrastructures.

(2) Consolidating and developing education and training programmes for designing, constructing and operating nuclear reactors at all levels.

6. The proper development of nuclear energy requires competent and independent safety authorities.

(1) They can rely on technical support organisations (TSOs), which equally must be competent and independent.

(2) There must be a technical dialogue between these authorities, operators, and vendors to ensure that safety requirements take into account the evolution of scientific and technical knowledge, and are based on an objective assessment of risks.

(3) The convergence of safety requirements among state regulators is a prerequisite for the standardisation of reactor models. This standardisation would in itself improve safety. In the present situation, safety authorities differ in the definition of the level of safety required, some implementing a cost-benefit approach, which is rejected by others; harmonisation is essential. The Academies recommend a "Risk Informed" approach which balances new safety requirements with their benefits.

7. The main challenges of the nuclear industry are ①waste management, ②control of costs, and financing of new projects, and ③conveying the message that security improvements of nuclear operations have attained such a high level that social acceptance should follow suit.

(1) Low and medium level waste management is being implemented industrially with proven and accepted solutions. High level-long lived waste (HL-LLW) conditioning technologies (with or without reprocessing of the spent fuel) exist. The scientific and technical community has not identified any major obstacle for the disposal of the conditioned HL-LLW in carefully selected deep geological layers. The characterisation of disposal sites needs to be pursued.

(2) The nuclear industry must intensify its efforts to control the cost and duration of nuclear projects.

A non-complacent analysis of the difficulties faced by some recent projects must be undertaken by the owners and vendors; the Academies recommend that the conclusions of these analyses be made public.

The Academies recommend that the nuclear industry accelerate the implementation of digital technologies, the use of which proves very beneficial in other industries. These technologies have helped to significantly reduce project costs and delays, while improving quality; three areas are of special importance:

①Use and further development of simulation models and their coupling with design;

②Digitalisation of the instrumentation and control systems;

③Use ofa digital platform (product lifecycle management, PLM) with a unique data base encompassing 3D-design, construction, operation and lifecycle management.

The introduction of these technologies requires cooperation with the safety authorities, and harmonisation of regulations, particularly in the area of cyber security.

Nuclear power is an industry with long lifecycles and requires significant up-front investment. It is important for it to receive adequate funding by international investment banks.

(3) The nuclear industry cannotdevelop if it is not socially accepted, and supported by Governments.

Safety aspects must be fully addressed and current and past improvements of nuclear operations clearly explained to the public.

Since the energy debate addresses complex technical and economic matters it can only be handled with well-structured information and reliable data.

The need for full and objective information is paramount. Non-governmental associations must be listened to, but so must operators, designers, safety authorities and experts in nuclear engineering and in economy. Governments should request expert opinions, including the expertise of their national Academies.

8. Nuclear industry could be shared with emerging countries through international and bilateral cooperation. Countries that have built a nuclear infrastructure, such as China and France, should help politically stable emerging countries to develop nuclear

energy in a safe and effective way. In parallel, the nuclear community should foster this development by making progress in two directions:

(1) Standardisation and stabilisation of the licensing and regulatory processes;

(2) Generalisation of long-term electric power purchase agreements (PPAs), with governmental guarantees.

Chapter 1 Introduction: The necessity of nuclear power in the future energy mix and challenges to overcome

As a source of social and economic development, energy is essential to human life, health and welfare. Meeting world energy demand, and reducing global emissions of greenhouse gases (GHG) raises fundamental challenges for the future of the planet.

Since the 1950s, the peaceful use of nuclear power has become one possible option for providing energy without resorting to fossil fuels like coal, oil or gas. In recent years nuclear energy has also been considered as a sustainable and reliable low-carbon source of energy, that guarantees energy supply, allowing global economic development while reducing emissions of greenhouse gases. Although nuclear energy has become one of the three pillars of global electricity supply its development traverses a critical period. The Fukushima nuclear unclear accident in Japan has had an enormous impact, raising public concerns about safety. This and some other reasons have led a few countries to abandon nuclear energy and terminate operation of existing nuclear plants. Return of experience and lessons learnt from the Fukushima nuclear accident have led to important safety enhancement plans of existing nuclear reactors and to further improvements of future installations so that the development of nuclear energy is being continued in many countries.

To obtain a noticeable reduction in greenhouse gas emissions in relation with climate change, diminish pollutant emissions of nitric oxides, unburned hydrocarbons and particulates, improve the atmospheric environment in the medium and long term, and achieve a sustainable development of human beings, requires a global energy system that will better respect the environment and that will use minimal amounts of fossil fuels. In the present situation where much of the electrical energy production is based on fossil fuels and more specifically on coal, nuclear power constitutes one of the most realistic options for supplying energy in a safe, efficient and clean way and simultaneously solving environmental and climate change problems.

Because nuclear power is a stable source of energy, it can reliably provide base load electricity and complements renewable energy sources such as wind or solar PV, which are mostly intermittent and are not easy to dispatch to respond to demand. In this respect, it is generally considered and supported by recent experience that the total share of intermittent renewable energies in the electricity mix in most countries, cannot exceed 30% to 40% without inducing unacceptable costs of electricity and leading to an increase in greenhouse gas emissions or a risk to the security of supply of electricity. The main cause of this limitation is the unavailability of electric energy storage, for which there is today no sign of a coming breakthrough. As it uses large concentrated power plants and requires lesser amounts of land in comparison with more dilute energy sources, it is capable of delivering the energy needed by the present megacities and by those that will arise in the future.

However, the development of nuclear power still raises many challenges and issues with regard to safety, management of radioactive waste, development and deployment of advanced nuclear energy systems, economics, public acceptance, etc. As two of the main countries with large capacities of nuclear power plants (NPPs), both China and France attach great importance to the peaceful use of nuclear energy in the world, and have the responsibility and willingness to help emerging nations in their development of NPPs and in their wish to resolve the challenges they will face.

In the continuation of COP21 and COP22 aimed at significant worldwide reduction of greenhouse gas emissions, the three Academies (Chinese Academy of Engineering, National Academy of Technologies of France and French Academy of Sciences) believe that their initiative to shed light on some of the complex issues related to nuclear electricity generation could send a strong and valuable message to academies in other countries, decision makers and society in general.

The present report reflects positions of the three Academies acting as independent bodies, and shall not be construed as positions of industrial actors in the nuclear power plant field or positions of either the French or Chinese governments.

In this report, the contributing academies aim at outlining the history and perspectives of nuclear energy, and address the key issues to be considered in order to make nuclear energy even safer and affordable for the benefit of developed and emerging countries.

Although this report touches on many issues, it is not intended to be comprehensive. It syntheses reflections and discussions carried out over a period of six months (Report preparation and discussion period from November 2016 to April 2017).

The present report comprises a set of seventeen chapters and a synthesis. Chapter 2 and 3 give a brief account of the history, problems and challenges of nuclear development and address issues concerning the deployment of Gen-Ⅲ NPPs. Chapter 4 to 6 deal with scientific aspects including promises and challenges of future reactor designs and specifically consider the Gen-Ⅳ situation and new small modular reactor (SMR) concepts. Technological issues are addressed in Chapter 7 and 8, include the safety issues and the need for technology support organisations (TSO), advances and challenges in digitalisation and in novel design tools. The importance of nuclear research, facilities and infrastructures is underlined in Chapter 9.

Chapter 10 is concerned with the question of education and training of manpower. One important aspect is that of attracting young graduates from higher education in the nuclear industry. Another is the training of employees in the utilities to give them the proper scientific background and the culture of safety management. In that respect, the role of simulators is underlined.

Chapter 11 and 12 deal with engineering issues including that of managing nuclear projects, questions of handling safety requirements while controlling costs and complexity. Chapter 13 considers the pertinence of international support of nuclear projects in emerging countries.

Chapter 14 to 17 deal with societal issues. An assessment is provided of the impact of global nuclear activities on human health during the last fifty years. Since safety is a central issue in the operation of nuclear power plants, it is necessary to examine (Chapter 15) the perception of risk and its relation with real hazards. Chapter 16 underlines the need to improve public awareness and considers the governance needed for that purpose. The reasoning pursued in Chapter 17 is devoted to the organisation and roles of the different stakeholders. This chapter also discusses the need to improve public understanding to limit the risks of increasing regulations that would complicate the development and operation of nuclear facilities.

Chapter 2   History, current situation, problems and challenges of nuclear development

---

Recommendations

Some major recommendations for allowing nuclear energy to play an effective role as source of low-carbon energy and as a part of the future energy mix in various countries can be made at this stage:

- Industry should make great efforts to keep nuclear projects in-line with their planned schedule and costs if public confidence in the capacity of nuclear players to master complexity is to be restored. To reach this objective, it seems necessary to explicitly understand the reasons of the poor performance of the recent construction of several Gen-Ⅲ+nuclear power plants in some Western countries, while at the same time identical construction in China is progressing according to schedule. This analysis should involve the stakeholders of these projects: owners, regulatory organisations, designers, engineering teams, contractors and subcontractors.

- Efforts should be made to bring SMR designs to the market as soon as possible to enhance the vision of a more flexible nuclear capacity that can also be used in regions with limited capacity grids or in emerging countries.

- Mechanisms for financing nuclear projects need to be re-examined jointly by governments of nuclear countries and by financial market players.

Civil nuclear energy represents one of the greatest technological challenges mankind has ever had to face because it features an unequalled number of facets all requiring a high level of knowledge and demanding approaches such as tolerant failure modes: safety, design, construction, commissioning, project management, legal and regulatory issues, waste management and decommissioning, financing, security and non-proliferation issues, health physics, environmental impacts, public acceptance, etc.

The complexity of nuclear engineering did not stop increasing since its origin in the 1940s and 1950s when science and technology, and the industrial capacity to build the "designed objects" were the only challenges to overcome.

Since that time, the mainstream of civil nuclear energy has progressed through a steady evolution of the technological reactor designs and of other supporting nuclear facilities: research reactors, laboratories, fresh fuel manufacturing and reprocessing facilities. Three generations of nuclear power plants have appeared successively with some overlapping without any major difficulty regarding the designs, each of them increasing safety by a factor of at least 10.

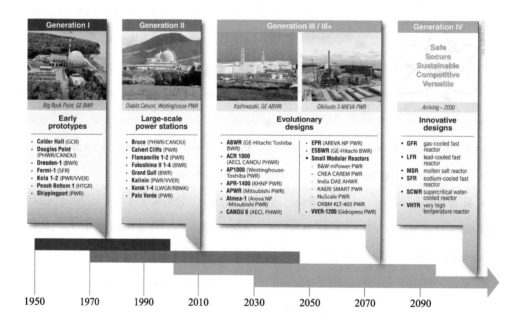

Figure 1 The successive generations of nuclear power plants. Reproduced from the Gen-Ⅳ International Forum (www.gen-Ⅳ.org)-Not shown in this diagram are the following Gen-Ⅲ/Ⅲ + Chinese Reactors: HPR1000 (CNNC PWR), CAP1400 (SNPTC PWR) and ACP100 (CNNC SMR PWR)

As a result of lessons learned from three major nuclear accidents (TMI, Chernobyl, Fukushima) and the "9 • 11" terrorist attacks, Gen-Ⅲ+must satisfy high expectations and standards on nuclear and radiation safety. This has led to cost increases that have reduced the competitiveness of nuclear energy.

The fourth generation is already under preparation and its deployment is expected after 2030. It specifically addresses the economics of fissile nuclear materials and the question of closing the fuel cycle.

In Western countries, the period from about 2003 to 2016 of the above-mentioned mainstream has been preceded by a kind of freezing of the nuclear sector, following the Chernobyl accident, when virtually no new construction project was launched, accompanied by a limited renewal of technical staff. This has not been without serious consequences for the development of Gen-Ⅲ and Ⅲ+projects that combine notably enhanced safety features with an increased safety design and licensing complexity.

Fortunately, two factors came into play to mitigate this difficult conjunction:

● Investments made by some vendors geared at modularity of construction and reactivation of reliable supply chains;

● The arrival of new clients/vendors such as South Korea and China having a huge potential of a trained and skilled nuclear workforce with "fresh eyes" and enthusiasm.

Inside the mainstream, decade after decade, the pressurised water reactors have commercially outperformed their competitors, first marginalising the heavy water reactors and then, after the Fukushima nuclear accident, the boiling water reactors, despite specific strengths of these two designs.

Much like the mainstream reactors, the design and construction of fuel cycle facilities and more generally the front-end and back-end industries did not, until today, experience any major difficulty and have not seen fundamental changes in their operation, except for the introduction of centrifugation technologies in enrichment facilities. One must also mention that industrial processes and technologies for partitioning/transmuting major actinides but also minor actinides have been validated at the R&D level; their deployment at industrial scale is only considered in conjunction with Gen-Ⅳ dedicated reactors or accelerator driven systems.

Current problems and challenges are mainly linked to the specific environment of nuclear projects. They can be summarised by a set of three issues, each one having serious and direct consequences slowing-down nuclear development and strongly interacting with each other, making it impossible to solve one without dealing with the other two! These issues are:

● Diminishing public confidence, nearly everywhere in the world, for large projects but especially nuclear ones, and consequently a demand for tougher safety requirements;

● A trend in many countries to ever increasing complexity of licensing and control processes;

● A growing difficulty to find public and private financing for initial nuclear investments, due to increasing costs and increasingly prudent banking rules.

With respect to these issues, some decision-makers, especially in Europe, consider that only government-owned vendors or utilities are able to handle nuclear activities. One might wonder if this is the best configuration for finding solutions for the transition to a de-carbonated energy system or if a competition between private actors might not yield a more optimal result in terms of safety and cost.

A crucial technological issue remaining partially pending is the long-term management of high-level long-lived nuclear waste. This has been a subject of considerable attention in various countries. On the basis of research carried out in France within the framework of two laws (1991 and 2006), it is believed that solutions exist that should help reducing the amount of waste. Such solutions stipulate the construction and operation of fast neutrons reactors, which could transmute minor actinides into fission products, the lifetime of which is about 300 years; far less than those of minor actinides. It is thus possible to be more optimistic about future nuclear waste, once reprocessed. However, at this point in time the most realistic solution is to dispose of existing waste or spent fuel in deep geological repositories. The few examples of such repositories under construction raise only few technological problems and those that remain are in the process of being resolved, but they also give rise to issues of public acceptance that need considerable attention.

A key to improving future public confidence in nuclear power will be to demonstrate the feasibility of decommissioning Gen-I and-II nuclear plants and other nuclear facilities in an increasing number of countries; increasing competencies on the technological and safety levels are certainly required, but even more important for public confidence will be the construction of nuclear projects on-time and within foreseen budgets.

Conclusions

It is of foremost importance to explain to decision-makers and to the public that nuclear energy effectively contributes to the limitation of global warming and that this is one of the most effective ways to provide abundant, safe and reliable base load electricity with demonstrated results.

It is similarly important to compare safety levels reached by nuclear activities with other human activities that imply a specific level of risk and explain how, in the nuclear domain, that risk has been reduced to a very large extent. Safety engineering with Gen-III and post-Fukushima enhanced Gen-II nuclear power plants has eliminated the need for countermeasures in case of severe accidents. In other words, these reactors will not entail major radiological consequences in case of severe accidents as all possible discharges will be contained within the reactor building.

It is worth recalling that most western Gen-II reactors will come to the end of their life somewhere before mid-century and that this will require a renewal of the nuclear power plant fleet.

Chapter 3　The feasibility and challenge for near and mid-term deployment of Gen-Ⅲ nuclear power plants

---

Recommendations

The economic competitiveness of Gen-Ⅲ/Ⅲ + power plants must be further enhanced through optimising reactor designs including digitalisation, modular and standardised construction, as well as innovative financial solutions and more supportive regulations. The French standardisation strategy of Gen-Ⅱ + reactors contributes to safety enhancement while reducing the construction cost. China has made great strides in digitalisation of design and engineering and in nuclear project management. Sharing the experience gained by these two countries would be beneficial for the future nuclear industry worldwide.

The current nuclear reactors connected to the grid in the world mainly belong to 2nd generation (Gen-Ⅱ) and improved 2nd generation (Gen-Ⅱ +) reactors. Their basic design was completed before the TMI-2 accident occurred; however, most of them were back-fitted with additional safety features, taking into account lessons learned from this accident (reactor safety valves; man-machine interface, etc.) and also from the Chernobyl accident (in-containment hydrogen recombiners, filtered containment venting systems). Before and after the Fukushima nuclear accident, Gen-Ⅱ reactors were also back-fitted to take into account station blackout (SBO) with additional passive or active systems, and loss of ultimate heat sink. It is worth noting that those back-fits were essentially not implemented in Japan; on the other hand, when back-fitted, these second-generation reactors can be qualified as Gen-Ⅱ+, and their safety level can be considered to be close to that obtained by later designs.

In the early 1990s, starting in Germany and France, safety authorities required that newly built reactors should meet the following safety objectives with respect to internal events:

● Accidents with core melt that would lead to early or large releases of radioactivity had to be practically eliminated.

● For severe accidents where the possibility of core melt could not be totally eliminated, design provisions had to be taken so that only limited protective measures in area and time were needed for the public (no permanent relocation, no need for emergency evacuation outside the immediate vicinity of the plant, limited sheltering, no long-term restrictions in food consumption) and provision had to be made that sufficient time was available to implement these measures.

● INSAG, a high-level advisory group to the IAEA general director, recommended fairly similar safety objectives in its INSAG-12 document (1999), well before the Fukushima accident[①]. Although IAEA *Fundamental Safety Principles*[②] did not include such requirements, they became part of the IAEA Safety Standard *Safety of Nuclear Power Plants: Design*[③], in line with WENRA safety objectives[④].

With respect to external events, regulators now tend to require consideration of an intentional airplane crash[④], with the need to demonstrate that safe shutdown can be achieved. Also, beyond design external hazards (earthquake, flooding), have to be considered, in order to demonstrate that no cliff edge effect would greatly impair nuclear safety.

In addition to safety improvements back-fitted in Gen-Ⅱ+ reactors, Gen-Ⅲ+

---

① INSAG. Basic Safety Principles for Nuclear Power Plants-A 1999 revision to INSAG-3 Rev. 1 (1999).
② IAEA. Fundamental Safety Principles-Safety fundamentals-SF-1-2006.
③ IAEA. Safety of Nuclear Power Plants: Design-Specific Safety Requirements-SSR-2/1 (Rev. 1)-2016.
④ WENRA. Western European Nuclear Regulators' Association-Report on Safety of new NPP designs (2013).

reactors include devices to collect the corium in the reactor cavity (core catcher), or systems to prevent melting of the reactor pressure vessel, in-containment (or in-reactor) water storage capacity (IRWST) to cool the molten core and a set of backup power sources. Mitigation of an airplane crash such as that corresponding to the September 11, 2001 attack requires a double containment building with adequate cooling systems.

The combination of all these features leads to reducing the frequency of a high-pressure core melt by a factor far greater than ten with respect to Gen-II + reactors and to essentially control all safety functions in case of an accident. Human error is reduced by implementation of a long-term operator intervention strategy. As a result, radiological consequences of a severe accident are drastically diminished, so that the objectives of "no permanent relocation, no need for emergency evacuation outside the immediate vicinity of the plant, limited sheltering, no long-term restrictions in food consumption" are achieved.

From an operational standpoint, the neutron flux reaching the reactor vessel is reduced, which will allow substantial life extension. Finally, collective dose to workers is decreased by a convenient choice of materials and equipment, easy to clean and maintain. Gen-III reactor nuclear waste should be substantially reduced compared to Gen-II per TWh.

American and European Utilities have translated regulatory requirements into more practical specifications (URD in the USA, and EUR in Europe). Several designs have been developed that comply with these requirements such as AP600/1000, EPR, VVER1000/1200, HPR1000, CAP1400, APR1400, APWR1500, ABWR, ESBWR, etc.

Reactors meeting these requirements and qualified as Gen-III + are ready for industrial implementation. Two ABWR units have been under commercial operation since 1996. There are eight AP1000 units under construction or commissioning in China and the USA with the leading project in Sanmen, China; four EPR units in Finland, France, and China, the first of which is located in Taishan. China also started construction in 2015 of four HPR1000 units. The Gen-Ⅲ + would be the main fleet for near and mid-term massive deployment of nuclear power.

Due to the enhancement of safety and complexity of the first engineering implementations, nearly all the Gen-Ⅲ demonstration projects were delayed or postponed with over-budget and financial pressure due to increased cost, except so far, for the HPR1000.

Due to significant efforts devoted to safety, the economic competitiveness of Gen-Ⅲ/Ⅲ + power plants must be further enhanced by optimisation of reactor design, utilisation of modern information technology, bulk procurement, modular and standardised construction, global supply chain partnerships, and innovative financial solutions along with more supportive regulations and supervision to minimise the construction duration, improve the thermal efficiency and the reactor utilisation, and consequently the safety and competitiveness of nuclear power.

The French standardisation strategy of Gen-Ⅱ + reactors features some important contributions to safety enhancement while reducing the construction cost. This strategy could also be used to reduce Gen-Ⅲ cost and maintain electro-nuclear power as the most competitive of all sources of base load electricity. A plan to progressively replace part of the French fleet with Gen-Ⅲ reactors is under consideration.

China has made a great plan to develop nuclear power, and along with the 30-year continuous construction of NPPs, a strong infrastructure has been set up, including development of competencies in R&D, design and engineering, manufacture, construction and erection, project management, commissioning and operation capacity. On this basis, China envisages strategic cooperation with related parties.

Conclusions

Most nuclear reactors connected to the grid are of the second-generation type (Gen-Ⅱ) with a high percentage improved to Gen-Ⅱ +, i. e. back-fitted for satisfying more stringent safety requirements. Gen-Ⅲ + reactors include devices for collecting the corium in case of core melt and other safety features, which reduce the frequency of high-pressure core melt by a factor larger than 10. Human error tolerance is increased through an operator intervention strategy leaving sufficient time for action so that the population in the wider vicinity of the plant need not be evacuated or worry about food consumption. Similar progress has been made at the front of neutron flux to the reactor vessel and reduction of exposure to radiation of workers. Such reactors are now ready for industrial implementation. However, such safety enhancements have a price and have led to substantial delays and budget overruns and loss of competitiveness.

Chapter 4   Promises and challenges of new and innovative future reactor designs (Gen-Ⅳ reactors)

Recommendations

In order to be ready in time to launch commercial Gen-Ⅳ sodium-cooled fast reactors (SFRs) and very high temperature reactors (VHTRs), the Academies recommend constructing and operating, during the coming decades, Gen-Ⅳ technological demonstrators of such reactors, including the facilities for the associated fuel cycle. Launching and operating SFR requires reprocessing of spent fuel to multi-recycle plutonium and uranium. Thus, special emphasis should be given to SFR fuel cycle. According to present technologies, SFR are a promising technology to provide electricity on the basis of fissile/fertile nuclear material accumulated since the beginning of the hitherto fission-based nuclear energy. VHTR have the potential to provide high temperature heat for multipurpose industrial use. According to the present national energy strategies, deployment of Gen-Ⅳ SFR and VHTR are expected after 2030.

A central issue in electronuclear energy is to extract the maximum amount of energy from nuclear fuel while at the same time maximising safety. To do this, one uses fissile and fertile isotopes of uranium (U) and plutonium (Pu). A fertile isotope such as ^{238}U gives fissile isotope such as ^{239}Pu by nuclear processes induced by thermal or fast neutrons.

4. 1 Present situation

Today worldwide nuclear electricity is produced (with the exception of a few units) through thermal neutron reactor types (TNR) fuelled with natural U oxide enriched in ^{235}U (UO$_x$ fuel). The energy released from the fuel comes from ^{235}U (around 70%) and ^{239}Pu fission, the latter isotope being produced in situ in the fuel from ^{238}U.

Some TNR reactors are partially fuelled with a mixed U-Pu oxide (MOX fuel). Plutonium is recovered from spent UO$_x$ fuel by reprocessing and U is the depleted uranium that remains after natural uranium has been enriched. The energy in MOX fuel originates from Pu fission (90%). Recycling plutonium from the UO$_x$ spent fuel saves consumption of natural uranium and diminishes SWU (Separation Working Unit) operation, but Pu can be recycled easily only once in appropriate TNRs.

A TNR does not use the total potential nuclear energy of natural uranium because nearly all the ^{238}U isotopes remain in the spent fuel. On the other hand, for safety reasons, the burn-up of fuel in TNR is limited to typically 50 GW · d/t (GigaWatt day per ton) with a temperature of the coolant around 300 ℃.

The only way to fission ^{238}U and to transmute it into ^{239}Pu, with a good yield, is to use fast neutrons. Fast neutron reactors (FNRs) are designed to take advantage of such neutron properties. The energy originates from the fission of all isotopes of U and Pu. In FNR the burn-up of MOX fuel can reach values of up to 140 GW · d/t (or more) and the temperature of the coolant can reach 500 ℃.

At this point in time, only a few FNR units are connected to the grid in Russia.

4. 2 Major expected change

In 2002, the Gen-Ⅳ International Forum (GIF) initiated a joint research effort on the nuclear systems of the future. This has given rise to an active partnership between China, France, Korea, Japan, Russia, USA and the EU. The technical goals and related evaluation index were defined by GIF in six areas: sustainability, economics, safety and reliability, waste minimisation, proliferation resistance and physical protection. Six most promising nuclear systems were selected, two of them were gas (helium) cooled reactors, another two were liquid metal (sodium and lead alloy) Cooled Reactors, one was super-critical water-cooled reactor and the last one was a molten salt cooled reactor.

Among the previous selection, the advanced, MOX fuelled sodium-cooled fast reactor (SFR), was considered to come closest to commercial status during this century by nearly all the GIF partners. The VHTR (very high temperature gas cooled reactor) system is being actively developed by China, Japan and Korea following developments in the US, Germany and South Africa. France has a limited contribution to the development of VHTRs but had initiated research on the GFR (gas cooled fast reactor).

SFRs are mainly devoted to electricity production and benefit from more than 400 reactor-years of operating experience since 1951. The VHTR is primarily dedicated to the cogeneration of electricity and hydrogen production, the latter being extracted from water by using thermo-chemical, electro-chemical or hybrid processes. Its high outlet temperature makes it attractive also for the chemical, oil and iron industries. Operating experience with VHTRs has been gained since 1963.

Both SFRs and VHTRs have potential for inherent safety and are designed to exclude any release of radioactive fission products into the environment under all operating or accidental conditions.

France is actively developing a 600 MWe Gen-Ⅳ SFR, based on its experience with operating SFRs, on spent MOX fuel reprocessing and on Pu recycling. China has plans to develop a 600 MWe demonstration SFR based on the CEFR experience. Japan has also experience in operating SFRs. Russia has FNRs connected to the grid and long-term plans for further developments. The civil nuclear program of India, based up to now on thorium, includes the launching of a MOX fueled SFR.

China follows a detailed long-term R&D plan related to VHTRs.

4.3 Promises of Gen-Ⅳ SFR and VHTR

At present the Gen-Ⅳ SFR reactors look promising in many ways. One important aspect is that nearly all the isotopes of uranium, plutonium, and heavier nuclei submitted to a flux of fast neutrons can undergo fission. This means that the depleted uranium, as well as the uranium and plutonium present in the spent fuels, can be multi-recycled in a FNR. Thus, a FNR can generate its own MOX fuel provided the spent fuel is reprocessed shortly after being unloaded from the reactor. In SFR MOX fuel, 15% of ^{238}U (from depleted U (DU) after enrichment process or U from reprocessing) is converted into plutonium. The energy extracted from natural uranium with a FNR is only limited by the industrial possibilities of recycling the spent fuel. The resource in fissile materials is therefore nearly unlimited with a fast neutron technology.

High burn-up allows leaving the fuel in an SFR twice as long as in a TNR and the minor actinides content is lower. This allows more economical operation of SFRs than of TNRs and opens the perspective of better waste management. The high-level radioactive waste from SFRs contains less long-lived radio nuclides compared to those originating from TNRs. This point is addressed in the chapter devoted to nuclear waste management.

Various configurations of SFRs are conceivable as they may be designed to produce less, exactly as much, or more Pu as they use. In the iso-generation configuration, the amount of Pu in an SFR park remains stable, in the "burner" configuration the stockpile of Pu can be strongly reduced. That reduces proliferation and allows to safely plan for the end of nuclear power generation should this be decided at one point in the future. An SFR can also be designed to transmute transplutonium elements such as americium.

Before launching of a new generation of SFRs one has to resolve various scientific and technological issues in laying out such reactors and in mastering the corresponding closed fuel cycle.

According to the Gen-Ⅳ Roadmap updated in 2014, the VHTR can supply nuclear heat and electricity over a range of core outlet temperatures between 700 ℃ and 950 ℃, and potentially more than 1000 ℃ in the future. The core outlet temperature refers to helium temperature, i. e. primary coolant temperature while the temperature of water steam depends on specific design features. R&D on the VHTR focuses on the fuel and the fuel cycle, materials, hydrogen production, computational methods for simulation, validation and benchmarking, components and high-performance turbomachinery, system integration, and assessment.

Gen-Ⅳ commercial reactors should have at least the same level of safety as Gen-Ⅲ TNRs. This requires many innovations with respect to the components of classical FNR reactors. Technological demonstrator designs of such reactors are under progress in France and China according to the GIF roadmap. SFRs should be considered as part of any future development of nuclear energy. This technology has the potential for providing a nearly endless source of dispatchable electricity by transmuting/fissioning all uranium isotopes and not only the 0.7% of ^{235}U in natural uranium. SFRs provide also a tool to control the plutonium inventory and to transmute long-lived actinides, which could facilitate the management of long lived nuclear waste. France was a pioneer in this technology and is actively developing a half-scale technological demonstrator of a commercial reactor (Astrid-600 MWe) incorporating operation feedback of past reactors and important innovations. China successfully operates a prototype reactor (20 MWe) of this kind since 2011 and plans to start construction of its CFR600, also a half-scale demonstrator, in the very near term. Construction of the first commercial Chinese SFR could start as soon as 2035. France foresees the deployment of commercial SFR reactors later in this century.

Conclusions

In summary Gen-Ⅳ SFRs and VHTRs are the most promising reactors to support the future of nuclear energy producing respectively electricity and both electricity and high temperature fluids/gases. At the present stage of development, where electronuclear energy is produced mainly based on TNRs, the increasing production of electricity augments the ^{239}Pu stockpile worldwide, at a rate of around 75 t/a, as well as other nuclear materials. This increase of large amounts of Pu stockpile induces non-proliferation issue, disposal of it as spent fuel if spent fuel is considered as a waste, or use of man-made fissile material in TNRs if it is considered as a resource. The quantity of Pu still produced in many countries and in the years to come, offers the prospect of launching in a few decades enhanced safety Gen-Ⅳ SFRs, providing energy for hundreds of years. But this implies drastically changing the present reactor system, setting up a novel nuclear fuel cycle and focusing R&D on the long-term use of nuclear energy.

Appendix A

A.1　Gen-Ⅳ initiatives in France

Since 2010, the CEA is in charge of the Astrid project (advanced sodium technological reactor for industrial demonstration), a project consisting of two parts: (i) The reactor design; (ii) The design of installations to produce its first specific MOX fuel, and then to recycle Pu and to test the transmutation of Am. Astrid must also prove that burning of Pu is achievable. The project gathers many industrial partners from the nuclear and non-nuclear fields and benefits from 40 years of Phenix and Superphenix FNR operation.

The Astrid reactor will be a 600 MWe pool type SFR loaded with MOX fuel, a power level that qualifies this system as an industrial demonstrator. It is designed to have four loops in the primary circuit and one steam generator or one nitrogen gas generator, depending on the design version, for each sodium loop of the secondary circuit. The tertiary circuit is a single line water-steam system or a two lines nitrogen system. Steam and nitrogen gas parameters are around 15 MPa and 500 ℃. Two shutdown systems and one supplementary independent shutdown system are designed for reactivity control. Several decay heat removal systems are connected with the hot pool or the vessel.

Innovations in the safety area focus on the control of the core reactivity in cases of sodium loss, and the cooling and confinement of radioactivity in all circumstances. CEA has patented a new heterogeneous core with a near zero-negative void coefficient, a major advance in FNR technology. It needs specific MOX fuel sub-assemblies. None of the in-service SFRs feature such an intrinsic stability core. The main chemical risk of water-sodium reactions is under control thanks to improved steam generator (SG) design. Alternatively, a sodium-nitrogen exchanger, that eliminates any contact between sodium and water, is presently under testing. This exchanger associated with a nitrogen turbine, would be the second major innovation.

New detection methods of sodium leakage have been patented by the CEA. In the case of a complete core fusion the corium is recovered. The residual power removal is ensured by several passive systems. Finally, double containment buildings and air-tightness systems should prevent any release of radioactivity into the environment should an accident occur.

Innovations in the operability domain relate to new in-service inspection systems and repair of reactor components. Special systems for handling of sub-assemblies for refuelling are being designed to save time.

The process for fresh MOX fuel fabrication for the Astrid reactor is industrially operational. The main problem to overcome is the industrial reprocessing of MOX SFR spent fuel, highly concentrated in plutonium, with a short turn-over of a few years. Thousands of tons of UO_x spent fuel have been reprocessed in France.

Despite major advances in SFR design over the last twenty years, Astrid still needs to be improved with further R&D in the area of, for example, materials for reactor components and fuel cladding. It is of special interest to develop steels for cladding allowing an increase of the burn-up to a level of 200 GW • d/t.

The decision to launch Astrid construction will not be taken before 2024. The French strategy for future nuclear energy consists of setting up a strategic reserve of Pu through the storage of un-reprocessed MOX TNR spent fuel and of other nuclear materials, such as depleted uranium. This should allow the progressive launching, beyond 2050 in the best case, of a self-sustaining park of SFRs, with, possibly, the transmutation of the americium that they will generate.

It is also possible to launch TNRs using the fertile mono-isotopic ^{232}Th, with ^{235}U or Pu as fissile isotopes. Fissile ^{233}U is produced in the fuel by nuclear processes on the ^{232}Th. ^{233}U has attractive fissile properties and it can be, or could be, produced in TNR by irradiation of ^{232}Th. Some prototypes of Th reactors operated more or less successfully. India experiments a "Th cycle". A modern version of a fast neutron molten salts reactor (FRMSR) considered in France features a high operating temperature (600 to 700 ℃), where molten salts (Th, Li and Be fluorides) act simultaneously as fuel and coolant. This system produces small quantities of heavy actinides, but fission products must be extracted periodically from the fuel, through batch or online pyro-chemical processes. Extraction of ^{233}U is also needed to maintain a stable flux of fast neutrons. Such reactors can transmute actinides online.

Many problems still need to be resolved in such systems. Although the use of liquid fuel is attractive, it raises issues of cooling for residual power, corrosion, containment of radioactivity, radiological protection against high-energy gamma rays, and management of new waste types. The MSR has balanced advantages and disadvantages but its development is not foreseen in France for commercial electricity production, at least unless massive fertile nuclides would become massively necessary.

The contribution of France to the VHTR development has been limited to materials and to hydrogen production. The CEA made significant efforts from 2000 to 2008 to design a prototype of a GCR as part of the GIF studies. This Allegro project was planned as a 75 MWth power system, helium cooled at 75 bar 850 ℃, a start-up core with a MOX fuel (30% Pu in stainless steel cladded pins) and in the long-term a UPuC carbide core with the same content of plutonium in silicium carbide cladded pins. In 2005 the CEA gave priority to R&D for Astrid while Allegro projects continued as part of a European programme with a reduced power level of 40 MWth since qualified carbide fuel is not yet available and further R&D is needed on materials.

A. 2　Gen-Ⅳ initiatives in China

1. SFR

As the second step of the Chinese "TNR-FNR-Fusion Reactor" nuclear strategy, the main objective of FNR development is to meet the energy demand and mitigate the possible shortage of natural uranium resource. Another goal is to allow for transmutation of long-lived nuclides.

The demonstration SFR, namely CFR600, will be built in China before 2025. The purpose of the CFR600 is to demonstrate the closed fuel cycle and establish the standards and rules for large size SFRs.

Technical choices made for the CFR600 are based on the following objectives:

● Meet safety design criteria of national nuclear power safety design regulations revised after the Fukushima nuclear accident.

● Use inherent safety characteristics and the in-depth-defence principle. Try to fit the reliability and safety requirements of the Gen-Ⅳ system.

● Adopt mature and technically proven components. All the innovative design should be fully certified by different methods.

● Try to raise the economic target compared with CEFR and other SFR projects.

The CFR600 is a pool type fast reactor loaded with MOX fuel. Its thermal power is 1500 MWth, for an electric power of 600 MWe. There are two loops in the primary circuit and 8 modular steam generators for each loop of the secondary circuit. The tertiary circuit is a typical water-steam system installed with one turbine. Steam parameters are 14 MPa and 480 ℃. Two independent shutdown systems and a supplementary hydraulic shutdown system are designed for reactivity control, and a passive decay heat removal system is connected to the hot pool. The primary and secondary containments are specifically designed for the reactor. The preliminary design of CFR600 was completed in 2016. The first concrete drop (FCD) is planned at the end of this year.

2. VHTR

The R&D Chinese programme for the high-temperature gas-cooled reactor (HTGR) began in the mid-1970s, and the construction of the HTR-10 test reactor was achieved in the 1990s. China is now moving forward to develop the high-temperature gas-cooled reactor pebble-bed module-HTR-PM-demonstrator project as a technical leader in the industry. In February 2008, the 200 MWe HTR-PM demonstration plant was approved as part of the National Major Science and Technology Projects. According to the roadmap report of the project, the prospects for HTR-PM development in China are its potential for highly efficient nuclear power technology as a supplement to pressurised water reactor (PWR) technology, in particular in the area of nuclear process heat. HTR-PM development will also contribute more generally through innovation in advanced nuclear technologies. The HTR-PM consists of two pebble-bed reactor modules coupled with a 210 MWe steam turbine. The helium temperatures at the reactor core inlet/outlet are 250 ℃/750 ℃, and steam at 13.25 MPa/567 ℃ is produced at the steam generator outlet. The first concrete of the HTR-PM demonstration power plant was poured on December 9, 2012, in Rongcheng, Shandong Province. The construction of the reactor building was completed in 2015 and two reactor pressure vessels were installed in 2016. In 2005, a prototype fuel-production facility was constructed at INET with a capacity of 100000 fuel elements per year. After that, a fuel-production factory with a capacity of 300000 elements has been constructed in Baotou, Northern China. Fuel production for the HTR-PM demonstration power plant started in 2016. Following operational demonstration, commercial deployment of HTR-PMs based on batch construction is foreseen, and units with more (e. g. six) modules are under design. It is also envisaged to develop units with multiple standardised reactor modules coupled to a single steam turbine.

Chapter 5　Promises and challenges of new innovative reactor concepts and technologies: Small modular reactors and advanced technologies

---

Recommendations

Light water reactors are a young technology which offers a huge potential for improvements.

● The development of small modular reactors should be encouraged and financially supported, as they offer a flexible means to address the requirements of a low carbon economy. International cooperation, in the framework of the IAEA, should be further enhanced, to develop new regulatory regimes adapted to their transportability from country to country.

● Promising results are expected from research and development, applicable to all Light Water Reactors. R & D funding of nuclear-specific technologies should be kept at a significant level. Transfer of technologies from other sectors should be systematically promoted. Following the initiatives by the IAEA and by the US NRC, regulators should review new features and technologies taking into account the experience gained with such technologies in other sectors.

In the context of the energy transition, market requirements have generated renewed interest in small modular reactors (SMR). They may find their place in a time of low-carbon economy. SMRs can become the source of stable and clean distributed energy.

As demonstrated by the development of SMRs, the Light Water Reactor technology is not frozen. Research and development is ongoing, and its results will be used to simplify nuclear reactors design and operation-whatever their size-while improving their safety.

5. 1 Small modular reactors

5. 1. 1 Research and development background

Since the 1970s, the industrial development of nuclear power plant based on conventional loop type reactor was accompanied by a continuous increase of reactors' generating capacity, up to more than 1500 MWe for the latest models, mainly for economic reasons.

Furthermore, research and development into innovative small or medium sized reactors (50~300 MWe) was also promoted for multi-purpose applications: isolated locations, small and medium-sized electricity grids not connected to neighbours; old coal-fired unit replacement; cogenerations to use waste heat in heating networks (district heating and process heating supply); seawater desalination; island development; progressive introduction of a nuclear electricity programme for a newcomer country.

It is obvious that despite often promising market and design studies, these SMRs have not yet been able to demonstrate their practicality for general application, mainly for slow development progress and economic reasons (installation, decentralisation and training costs), but also location and implementation delay.

5. 1. 2 Renewed interest

For a number of years, mainly at the initiative of the DOE in the USA, there has been renewed interest in SMRs on the account of advanced nuclear energy leadership and major ongoing developments in the energy landscape and nuclear technology. The SMR is a game changer which may provide a different nuclear power co-generation solution with high safety level.

On the energy side, four reasons can be mentioned: ①The need to reduce the use of fossil fuels; ②Decentralisation of electricity generation (renewable energies, smart networks, energy storage); ③The need for operators to be agile and flexible, and finally; ④The problem of financing long-term investments.

On the technology side, SMRs include a large variety of designs and technologies, such as: Integrated PWRs which can be employed at near term; Small-size Gen-Ⅳ reactors with non-water coolant/moderator which can be employed at medium-long term; converted or modified impact loop type SMRs, including barge mounted floating NPP and seabed-based reactors. Three major developments have repositioned the competitiveness and attractiveness of SMRs:

(1) The possible use of the concept of passive safety for smaller reactors, which satisfies increasing safety requirements and at the same time allows design simplification.

(2) The emergence of in-factory modular construction capacities which should reduce overall costs and building-time on location.

(3) Like the "plug and play" concept, the power station is built entirely in the factory, transported and connected to the grid; the only significant local operation is to connect the station to the electricity network.

Among the many SMRs concepts being studied in the USA, Russia, China, South Korea, Japan, UK, Argentina and also France, two generic models emerge: terrestrial and transportable SMRs.

● Terrestrial SMRs aim at nuclear island modularity and need to be installed at a specific location with civil engineering and additional ancillary facilities, turbine-generator unit, network connections.

● Transportable SMRs, completely decoupled from the operation-site, providing agility, flexibility and reversibility while at the same time reducing overall acquisition time to a minimum for a newcomer.

As a typical terrestrial SMR, ACP100 (Figure 2) is a 125 MWe small modular reactor developed by China National Nuclear Corporation (CNNC). ACP100 adopts proven and practical light water reactor technologies with five significant technical features: integral reactor, inherent safety features, fully passive safety system, underground arrangement, and twin unit sharing one 250 MWe turbine generator.

ACP100

Figure 2   ACP100: Small modular reactor developed by CNNC

Several transportable SMRs are being built or developed:

● A concept on a barge proposed by Russia and China. A first Russian unit is being finalised while China is rapidly developing ACP100S (125 MWe, Figure 3) and ACPR50S (50 MWe).

● A submerged concept Seanergie (Figure 4) studied in France capable of providing more power (160 MWe). An alternate onshore option is also studied.

Figure 3   ACP100S floating NPP developed by CNNC

Figure 4　Seanergie: An immersed SMR

5. 1. 3　SMR challenges and solution SMR

Industrialisation of those models requires proof of their competitiveness, public acceptability, and a regulatory framework for the transportable reactors that IAEA is working on.

To comply with actual market requirements, new SMRs have to be developed with truly innovative concepts and surely not as a downsizing project of present Gen-Ⅲ reactors.

The economic competitiveness of SMRs can obviously be improved by innovative design. Economically, SMRs can be competitive with intermittent wind power, solar power, gas power generation and diesel generators for particular applications.

If solutions similar to plug and play, with a design completely independent of the installation site, are confirmed, they may be best placed to fully comply with market requirements and thereby contribute to a realistic energy transition.

5. 2　Innovative technologies to be implemented in large reactors

Commercial nuclear technologies are only a few decades old. Because of stringent safety requirements, innovations are only slowly flowing into the reactor designs. Therefore, a huge potential of improvements is lying ahead, by either implementing technologies already applied to other sectors, or specifically developed for nuclear applications. Both SMRs and large commercial light water reactors will benefit from such technologies. Nuclear-specific technologies, or technologies transferred from other sectors.

1. Nuclear specific developments

(1) High performance fuels:

● Increased burn-ups with limited swelling and fission gas releases;

● Tolerant fuels withstanding higher temperature without melting, and preventing or limiting hydrogen generation in case of accident.

(2) Improved in-core instrumentation with better accuracy, allowing less conservatism in design analysis and operation.

(3) Improved understanding of corium behaviour, to optimise in-vessel fuel retention in case of accident.

(4) Implementation of up to date simulation methods, coupling thermohydraulic and neutronic calculations in real time substantially enhance design and operation capabilities.

2. Technologies transferred from other sectors

(1) Digitalisation of design, procurement, construction and project management of nuclear facilities (see Chapter 8 in this report).

(2) New composite materials to replace steel for low pressure circuits.

(3) Advanced concrete with high mechanical properties, and leak-tightness.

A high priority should be given to R&D in these fields; international cooperation should be encouraged, when Intellectual Property issues are not at stake.

Conclusions

Nuclear energy is a relatively young technology, with a huge potential of improvements, including Gen-Ⅳ reactors, and also small modular reactors. Technological bricks such as accident tolerant fuel would enhance safety and simplify the system in order to increase the competitiveness of all light water reactors. Highly innovative SMRs could offer new solutions to further develop flexibility and decentralised production. They also allow smooth ramp up in financing and in local nuclear skills development for newcomer countries and rapid access to nuclear electricity.

Chapter 6   Radioactive waste management status and outlook for the future

Recommendations

Safe management of radwaste (RW) is a key issue for expanding nuclear energy. All nuclear countries are faced with this problem and have to develop the administrative and technological tools to deal with the corresponding issues. Main hurdles are to be overcome in order to dispose of the ultimate long-lived RW. There is general agreement to site repositories for such RW in deep geological rocks. The Academies recommend to intensify scientific and technological research and development for managing all types of RW with special attention to the qualification of host rocks for disposal of high-level long-lived wastes. The Academies also consider that international cooperation on this topic needs to be promoted.

Safe management of radioactive waste (RW), which is produced in every stage of the nuclear fuel cycle, is an important contribution to demonstrating that nuclear energy can be managed in a safe and sound manner. This issue has a great significance in the peaceful, sustainable, and scalable development of nuclear energy and strengthening of public confidence. Radioactive waste is categorised according to activity and half-life of the main radionuclides contained therein. Another criterion is heat generation and the categorisation of RW depends on the classification system of individual countries according to their national radwaste policy. The management of medium and high activity long-lived RW is an ongoing and difficult task, due to the necessary heavy radiological protection measures and to heat generation from the most radioactive waste.

6. 1   Fuel cycle strategy and RW characteristics

Nuclear fuel cycles fall into two categories, a once-through fuel cycle (also designated as open cycle) and a closed fuel cycle, in which the spent fuel is reprocessed to recycle uranium (U) and plutonium (Pu). These cycles reflect political decisions on the disposal of Pu as waste or use as nuclear fuel material. In a once-through fuel cycle, the spent fuel will be disposed of, as an ultimate high-level RW, after a long-term temporary storage to allow thermal power decay of the sub-assemblies. Final repositories are nuclear installations constructed in deep geological rock formations. According to safety analysis, the host rock shall isolate and confine radioactivity over hundreds of thousands of years.

In a closed fuel cycle, spent fuel is reprocessed, yielding ultimate short and long-lived "processing" and "technological" RW. The long-lived processing RW, containing all the radionuclides present in the spent fuel, except U and Pu, is highly radioactive. All the long-lived spent fuel RW, initially stored in facilities, will be sent to a final deep geological repository as stipulated in the open cycle option. Most nuclear power countries, including France and China, have adopted for a closed fuel cycle. Indeed, the closed fuel cycle is the cornerstone of a sustainable nuclear energy system, where U and Pu are multi-recycled in FNR as France and China intend to do. The closed fuel cycle improves the efficiency of natural fissile resource consumption and reduces the toxicity of the ultimate waste since it does not contain Pu. However, Pu has to be managed in the fuel cycle until the end of the recycling process. The total volume of ultimate waste is slightly reduced.

6. 2   Progress on RW management

RW management involves controlling, collecting, sorting out and processing waste. These actions are then followed by conditioning different RW into appropriated packages to avoid dispersion of radionuclides. Packages are further transported to storage and are finally disposed of. The disposal step is regarded as the core objective and ultimate management target. The next chapters focus on the current situation of RW management (excluding very short-lived RW or RW emanating insignificant radiation doses that are subject to exemption and/or clearance practices, where they exist).

1. Very low-level waste (VLLW)

Whatever the radionuclides contained in VLLW, their activity is so lowthat this waste can be disposed of in surface installations like shallow landfills. These account for the largest volume of nuclear RW. Indeed, dismantling of nuclear facilities can give rise to large quantities of VLLW, with a proportion of 50% to 75% of all decommissioning RW. It can thus be predicted that the bulk of VLLW production will occur in the future. Above and beyond the positive development of volume reduction technology, the construction of large VLLW disposal facilities is imperative.

2. Low and intermediate level waste (LILW)

If only short-lived radionuclides are concerned, the final management issues of LILW can be solved by constructing and operating surface/sub-surface disposal facilities, including shallow trenches, near-surface concrete structures, underground rock caverns or tunnels, vertical large diameter boreholes. Practice and experience of LILW in waste conditioning, transportation of packages, reception on site, disposal of packages and safety assessment is available.

Low-level long-lived RW, produced in large quantities, is a special category of waste that is difficult to manage. If sub-surface disposal is selected it must be proved that long-lived radionuclides will remain isolated from the biosphere for a long time due to their long half-life and radiotoxicity.

3. High level waste (HLW)

HLW is a mixture of toxic and hazardous materials containing large amounts of long-lived radionuclides. Examples are sub-assemblies of spent fuel or packages of vitrified fission products and minor actinides. HLW has to be stored during decades for cooling before being transferred to a centralised deep geological repository. Countries around the world have made extensive efforts to select sites and design repositories for HLW. Belgium and France selected sites in clay layers, Finland and Sweden site their HLW waste disposal in relatively homogenous granites. Germany, USA, UK and many others countries are still in the process of selecting sites in an appropriate host rock, having so far abandoned salt domes and volcanic rocks. China is now studying both granite and clay rock for siting a HLW repository. The choice of a site takes several decades and requires experiments in Underground Research Laboratories (URL), which appear unavoidable. Up to now only Finland has obtained permission to go ahead with the construction of a HLW repository in 2015. Based on extensive research work on the selection of host rock and preliminary disposal design (including the sealing of the repository), one can expect that HLW will be isolated in safe ways.

Conclusions

The international management of radioactive waste has clear goals. It asks for the safe deployment of continuously improving solutions for the various types of waste (VLLW, LILW, and HLW). This requires experience in resolving connected conflicts through information transparency, public understanding of issues and participation. For the future development of RW management, the following points need to be considered.

1. National policy should pay permanent attention to management of radioactive waste issues

Decisions at the national level are the primary driving forces to carry a safe management programme relying on a firmly based legal system, well founded scientific studies and technical solutions, detailed and coordinated planning and adequate fund investment.

2. Research and development on science and technology of RW management should be intensified

RW management requires advanced science and technology in the treatment and disposal of waste to achieve safety. The construction of a deep geological repository constitutes a central issue and a considerable challenge. Each component and control system must be operational during more than a century and its closure must ensure that the waste is totally isolated. It is important to intensify research and development in the underlying scientific disciplines and encourage technological innovations to effectively reach breakthroughs in key issues of waste management.

3. International cooperation and broad prospects in the future should be promoted

The management of RW is an important issue that needs permanent attention by all those engaged in the nuclear industry. It is important to develop international cooperation and share knowledge, information and technology in this domain.

Appendix B

For each nuclear country, the characteristic features and trends of RW management depend on many factors. The main driving factor is the national choice on an open or a closed fuel cycle. Radioactive waste management in France is typical for a country aiming at a closed fuel cycle.

Today 90% of the RW is produced by the electronuclear industry in the operation of facilities dedicated to manufacturing, using, recycling and storing nuclear fuel. This figure is not expected to change, according to the French energy policy with regard to spent nuclear fuel. All the unloaded sub-assemblies of UO_x spent fuel of the current fleet (58 PWR-and 1 EPR-systems, 62 GWe, 420 TWh/a) will be reprocessed to recycle only once Pu and U in MOX and URE fuels. The sub-assemblies of spent MOX fuel will be stored as a strategic stockpile of Pu for the future. The figures correspond to a reactor life-time supposed to be 50 years. All other non-electronuclear spent fuels will also be reprocessed.

The French system identifies five families of RW according to the criteria of radioactivity levels and decay time of such waste, corresponding to a pragmatic view of management, corresponding to so called "radwaste channels". Table 1 presents the 2013 volumes and the expected total figures for 50 years of reactors life-time. Figure 5 illustrates the French electronuclear nearly-closed cycle. At each step, the size of the surface of the circle is proportional to the amount of RW produced in 2013.

Table 1　Radioactive waste volumes

Radioactive waste	Abbreviation	Volume up to year 2013/m³	Activity relative to total activity in/%	Total volume/m³ up to 50 years lifetime
High level-long lived waste	HL-LLW	3200	~98	10000
Intermediate level-long lived waste	IL-LLW	44000	~2	72000
Low level-long lived waste	LL-LLW	91000	0.01	180000
Low and intermediate level-short lived waste	LIL-SLW	880000	0.02	1900000
Very low-level waste	VLLW	440000	<0.000004	2200000
Total		1460000	100	4300000

Legacy-and dismantling RW of nuclear facilities are included in the total amount. The most active RW from reprocessing (5000 packages of HL-LLW and 170000 of IL-LLW), to be disposed of, do not contain Pu or U (except for losses from the Purex process). Furthermore, it should be noticed that, in comparison to an open fuel cycle, a closed fuel cycle with recycling of Pu leads to reducing the production of packages of HL-LLW by about 90% (0.8 against 8 m³/TWh) but to an increase of IL-LLW (0.6 against 0.08 m³/TWh).

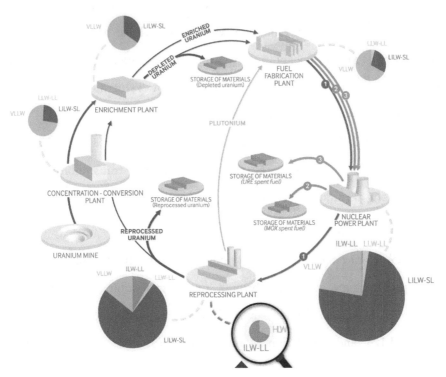

Figure 5　French electronuclear nearly-closed cycle (From Andra)

1: UO$_x$ fuel; 2: MOX fduel; 3: recycled Uranium. Dashed line: running and dismantling waste

The nextchapters focus only on the management of long-lived waste and VLLW, which are at the root of the main problems. Some indications will also be given on the management of other waste: LL-LLW, IL-SLW, uranium tailings and technologically enhanced naturally occurring radioactive materials (TENORM), which are solved or in progress.

1. French Institutional Framework

Andra is the nuclear operator in charge of the RW. ASN in addition to its control role provides safety guidelines for Andra and for radioactive waste producers such as CEA, EDF and Areva. The law No. 2006-739 of June 28, 2006 defines the French management policy of the various kinds of radioactive matter and RW and ① sets the research areas, milestones and targets to be reached, taking into account societal requirements; ②establishes a management plan (elaborated by ASN and the Ministry of Industry) including interim storage/disposal of RW and partitioning/transmutation of long lived radionuclides. It provides a coherent framework for the global management system and takes dispositions to continuously improve a safe management of RW; ③establishes a new National Assessment Board (CNE2), in charge of evaluating progress in research and studies dealing with the management of RW.

The 2006 law follows the law No. 91-1381 of December 31, 1991, which was devoted to research to be done regarding partitioning, transmutation, geological disposal, and long term storage.

2. Progress and challenges

Progress in RW management has been continuous, from conditioning to final disposal. Associated problems are addressed through long-term planning of R&D. One major achievement has been the selection, by Andra, of a site in clay for the disposal of HL-LLW and IL-LLW and the design of the facilities for a deep geological repository named Cigéo.

3. Cigéo repository for HL & IL-LLW

The 2006 law stipulates that the HL and IL-LLW disposal facility will be reversible for at least one century. It is considered that reversibility is the ability, for successive generations, either to continue the construction and further exploitation of the successive steps of storage, or to re-assess the choices made previously and to change the management scheme. This includes the possibility of recovering packages of waste already stored in a manner and for a period consistent with the strategy of operation and closure of storage.

Callovo-Oxfordian clay has been chosen as host bedrock. Cigéo will be launched quite soon at Bure, a location near the underground laboratory LMHM (Laboratoire de Meuse Haute-Marne) operated by Andra over a period of 15 years. The clay layer, 130-meter-thick, 500-meter-deep, and located in the east of the Paris basin, shows a remarkable lateral continuity and a good homogeneity in composition and structure. Investigations on regional and local geology, hydrogeology and geochemistry show no faults or connection with convective flow between the upper and lower aquifers through the clay. Clay pore water has a very long residence time. Water composition and diffusion coefficients of the fastest diffusive radionuclides have been measured in situ in the LMHM. The data show that clay is an efficient ultimate barrier to stop radionuclide migration up to a million years. Mechanical properties allow the building-up of underground facilities designed to separately dispose HL-LLW and IL-LLW in specific cells and to last more than a century during the operation of the repository. On site investigations and design of Cigéo began in 1991.

4. LL-LLW

The LL-LLW is formed by materials containing Ra and its daughters, large or small pieces of graphite containing ^{36}Cl and ^{14}C, packages of slugs in bitumen containing trace amounts of actinides. They are stored in various places waiting for disposal. Andra is looking for a dedicated sub-surface site in clay. Disposal of such waste in such a place is a challenge due to the long-lived radionuclides to be confined so close to the biosphere. R&D is conducted to see if processing of some LL-LLW could help in disposing of this category of waste. The LL-LLW will amount to around 200000 m^3.

5. LIL-SLW

The IL-SLW, mainly originating from electricity production $(1.9 \times 10^6 \ m^3)$, is and will be packaged and disposed of in the CSA (Centre de Stockage de l'Aube, Soulaines-Dhuys) centre, open since 1992 and foreseen to be in operation up to 2100 (capacity of $10^6 \ m^3$). The previous repository, CSM (Centre de Stockage de la Manche, Digulleville), open in 1969 $(0.5 \times 10^6 \ m^3)$ has been closed and is under monitoring since 1993.

6. VL-LLW

VL-LLW is currently disposed of in a special centre: the Cires (Centre Industriel de Regroupement et de Stockage, Morvilliers). Its authorised capacity is 650000 m^3. It could be extended to 900000 m^3 and should be full by 2030. In addition to the extension of capacity of Cires, another high-capacity centre needs to be opened to cope with the disposal of VL-LLW, originating from dismantling activities.

VL-LLW raises many problems due to the large foreseen quantity (estimated to be 2.2×10^6 m^3) because the French system of RW management defines no clearance level. For a few years, research organisations, industry and the authorities have been developing innovative methods for the management of dismantling-derived materials also classified as waste, although they contain little or no added radioactivity. In order to do so, one has to develop methods for measuring very low radioactivity levels in large batches of materials that could be used to support an innovative VLLW management strategy. Andra and waste producers are looking first to all the possibilities for drastically reducing VLLW production, for instance by recycling large streams of metallic components. Concerns about recycling are numerous and require at least the definition of thresholds to be reached in decontamination before reuse.

7. Technologically enhanced naturally occurring radioactive materials (TENORM)

Up to now Tenorm waste, originating from non-nuclear industries were managed in technical centres for industrial waste, equipped to detect radioactivity above fixed thresholds. This management is under reconsideration, particularly with regard to their quantities and their activity. This raises concerns similar to those pertaining to LLW.

8. Other waste

Uranium has been mined in France. About 50 Mt of U radwaste (tailings) are disposed of *in-situ*, in 17 sites, near the place of their production in old open-air mines, basins or bottom valley closed by dams. The sites are monitored and submitted to specific regulations to take care of radon migration and daughter products. A large amount of radwaste from processing of yellow cake, which are more or less similar to mining radwaste, is stored in Malvesi.

In conclusion, France has set up under the control of ASN a safe radioactive waste management system to support its nuclear activities according to the strategy of reprocessing/recycling all spent fuel. The Cigéo project of a deep geological disposal will be launched in a few years' time. Management of VLLW is aimed at reducing the large volumes resulting from dismantling of reactors and nuclear facilities and at defining a new high capacity disposal site. From several debates between all stakeholders, including the public, it has become clear that people want to be "secured" and not "reassured". The major question is: how can one have confidence in the management of such long-lived radwaste. Reversibility of disposal of radwaste, as stipulated by the law, reconciles ethics and transmission of duties to next generations. It helps making public acceptance a reality. The next step will be to set up the governance of repository reversibility, which by law must involve the public.

Chapter 7　Technical support organisations related to safety

Recommendations

- Technical expertise of regulators, either built in-house or provided by TSO is essential to secure a high level of nuclear safety.

- Independence and transparency of TSOs should be kept at the highest possible level, and not be compromised.

- Competences of TSOs should be ever increased on a continuous basis, for them to be able to assess new, multiple, and innovative technologies, including Gen-IV reactors.

- International cooperation between TSOs should be strengthened to better harmonise requirements, and assessment methodologies and criteria.

- TSOs of nuclear-developed countries should assist and support emerging countries, to build up their national safety expertise.

All nuclear related activities are deeply regulated (Ref. IAEA *General Safety Requirement-Governmental*, *Legal and Regulatory Framework for Safety-Series* No. GSR Part 1 (Rev. 1)).

Throughout the lifetime of a nuclear facility, the owner/operator has the prime responsibility for safety; but the regulator shall be fully empowered to control the owner/operator.

Beyond the administrative and legal tasks of issuing a construction, first fuel loading, operation or dismantling license for any specific facility, the regulator needs to be supported by a strong technical expertise/organisation. Such expertise/organisation will be used to develop safety rules, regulations, and requirements in a coherent fashion, to assess and evaluate the licensees' compliances and non-compliances.

Many organisational schemes exist for a regulator to access nuclear expertise. It can be built in-house by the regulator, or contracted to external parties. And when safety expertise is subcontracted, a preferred specialised agency is often considered, referred to as the technical support organisation (TSO-also the technical and scientific support organisation).

In the US, the NRC owns the expertise to develop nuclear safety regulations and regulatory guides, and to assess compliance with its very prescriptive requirements. In France, the regulator (Autorité de Sûreté Nucléaire, ASN) develops general guidelines, and owns a strong expertise in the field of integrity of the primary and secondary circuits, supplemented with contracts to universities and research institutes in this specific field; but relies on the Institut de Radioprotection et de Sureté Nucléaire (IRSN) for specialised nuclear engineering expertise. IRSN is a spin-off of the former safety division of the French Atomic Energy Commission, later merged with the radiophysics division of the Ministry of Health. Essentially all technical safety assessments, among other tasks, are assigned by ASN to IRSN, the French TSO.

In China, the regulator (National Nuclear Safety Administration, NNSA) issues nuclear safety rules and guides, prepares and promulgates nuclear safety regulation, controls their implementation, establishes principles and policies of nuclear safety. All civil nuclear facilities are regulated by NNSA. China Atomic Energy Agency (CAEA) is in charge of national nuclear emergency.

Most of the nuclear expertise is accessed by NNSA from several TSOs: ①Nuclear and Radiation Safety Centre (NSC) of Ministry of Environmental Protection which provides most of technical support work needed by NNSA. ②Radiation Monitoring Technical Centre (RMTC) of Ministry of Environmental Protection; its main duty is to provide national radiation monitoring support. ③ Suzhou Nuclear Safety Centre

(SZNSC) which provides technical support in the field of quality assurance, mechanical equipment and research reactors. ④ Nuclear Equipment Safety and Reliability Centre (NESRC), which participates in reviewing nuclear pressure retaining components. ⑤Beijing Review Centre of Nuclear Safety (BRCNS), the first founded TSO in Beijing Institute of Nuclear Engineering (BINE); its main duty is to provide technical support in the field of nuclear engineering not designed by BINE. All safety assessments are subcontracted to these TSOs, and supplemented with contracts to universities and research institutes in specific fields, such as Tsinghua University, China Institute of Atomic Energy, China Institute for Radiation Protection, etc.

When reviewing international experience in this field, there is no evidence for claiming that any organisational model is superior to others; many models work, provided they ① fit the technical history and administrative organisation of each country, ②channel the best available expertise to the nuclear Regulator and ③foster the development of such expertise.

However, a few basic principles generally applicable to nuclear safety expertise can be drawn up, inter alia:

(1) Independence: the expertise body (TSO) shall not undertake work likely to compromise its neutrality or likely to lead it to assess its own work. If the expertise body works for several clients, including the safety authority, potential conflicts of interest should be identified and prevented.

(2) Transparency: the TSO should be organised to enhance independence and transparency of its operation, including full disclosure of its reports (except when serious confidentiality issues are at stake).

(3) Quality assurance: the expertise body should comply with a stringent quality assurance programme, ensuring the assessment of the competence of its experts, the traceability of its expertise, from task acceptance to provision of deliverables.

The expertise body can provide training services, and manage its own R&D programmes, which are means to enhance the qualification of its staff.

TSOs should not work in an isolated manner. Although nuclear safety is a sovereign issue for any state, experience and best practices should be shared. Owners/operators are sharing their experience on a daily basis in the frame of the World Association of Nuclear Operators (WANO), and organise peer reviews on a regular basis. The IAEA and OECD provide fora for safety authorities to meet and cooperate, in the frame of working groups or peer-reviews as requested by the Nuclear Safety Convention. No such forum exists for TSOs; however, major TSOs in the world took the initiative under an IAEA umbrella to hold regular meetings, the 3rd TSO was taken place in Beijing in 2014; such exchanges should be encouraged. Although safety requirements are left to each state, sharing experience and expertise between TSOs can only benefit each of them.

The development of nuclear energy in emerging countries raises specific issues with respect to their access to safety expertise. When nuclear technology is imported, a "reference plant" is generally considered. Clients should be given access to safety expertise from the origin country of this technology, and to the safety case of the reference plant, provided the requirements of independence and transparency are complied with. Furthermore, TSOs of more advanced countries should help and assist newcomers to set up and develop their own safety expertise, to the maximum extent possible.

When looking back to more than fifty years of commercial nuclear development, the invaluable role of TSOs cannot be overestimated. Analysis of nuclear incidents or accidents demonstrates that the absence of an independent and transparent TSO, implementing strict Quality Assurance programmes has had serious consequences. In the future TSOs should continue serving the global nuclear community. Some key challenges lying ahead are:

● Keep and increase their competences to be able to assess new, multiple, and innovative technologies, including Gen-IV reactors.

● Strengthen international cooperation to better harmonise requirements, and assessment methodologies and criteria.

● Support safety authorities and TSOs of emerging countries, to build up their national safety expertise.

Conclusions

Nuclear regulators need to rely on a strong technical expertise, which they can build in-house, or contract from a TSOs; there is no evidence that any organisational model is superior to others.

TSOs should comply with stringent requirements of independence, neutrality, transparency, and quality. While they would largely benefit from international cooperation and exchange of experience, there is, as yet, no dedicated forum for cooperation and peer review. Notwithstanding, TSOs of advanced nuclear countries can and should play an important role in assisting newcomers to develop their own safety expertise.

Chapter 8   Challenges for the future, including digitalisation and novel design methodologies

---

Recommendations

The nuclear industry shall take full benefit of digitalisation at all stages of its development.

● More accurate design codes simulating reactor operation in normal and accidental conditions can take advantage from progress in IT including fast increase in the speed and storage capacity of computers. They pave the way for implementing new methodologies that need to be licensed through positive interactions with safety regulators.

● Digitalisation of instrumentation and control systems is a reality; but safety requirements are not harmonised between regulators, leading to different solutions to fulfil the same functionality, depending on the country. Regulations and rules coping with new threats, like cyber security are also not harmonised and progress in these fields should be made at the international level.

● Many industries take full advantage of new digital technologies for design and project management (CAD tools, project life management). The nuclear industry will greatly benefit from these tools. Sharing of experience in this field is encouraged.

Nuclear energy must deliver low carbon and competitive electricity by reducing cost and construction time of new nuclear power plants, without affecting safety, and by better controlling outage times of existing plants.

To cope with these challenges, nuclear energy needs to make the best use of the available tools stemming from digital technologies, and above all "Industry 4. 0", PLM (product lifecycle management), 3D scan for numerical twin of an existing facility, big data and internet of things, virtual and augmented reality, 3D printing, etc.

Furthermore, nuclear energy is developing in an open and global economy, with focus to "One belt and one road" in China, and open competition in Europe.

Digital transformation of nuclear engineering will help achieving competitiveness and provide economic gains, while improving safety; it should also significantly enhance and improve efficiency and benefit the working organisations, which must adapt to digitalisation.

At stake:

● Increased control of construction schedules of new nuclear plants;

● Increased availability factors of existing plants, by better planning, scheduling and controlling outages;

● Lifetime extension of installed base;

● Increased competitiveness of 3rd generation nuclear power plants.

For the installed base, which has not benefited so much from a numerical model at the design stage and updated during operation, creating a "numerical twin" thanks to a 3D scan, and sharing it with subcontractors, will allow time-saving for maintenance operation, testing accessibility for maintenance, and facilitate training of operators, up to the dismantling phase. The numerical model also provides tools to minimise workers' exposure to radiations.

For new projects, considering the long lifetime of a nuclear project (typically 60 years) and the very high level of safety requirements, it is critical to use data management tools for managing in real time the reactor configurations, covering specifications, design, construction, operation and dismantling.

Such tools as for example PLM (product lifecycle management) exist and are routinely employed in the aerospace sector.

Lack of complete control over data is a major source of delays during construction phase and cost-overruns due to rework.

For a successful implementation of a PLM numerical platform, a change in industrial organisation is mandatory. This requires:

- System engineering with "architects" (chief engineers) fully responsible for the whole system (nuclear island, conventional island, pumping station, ⋯) from design to construction and commissioning.

- A strict data governance, with a process to authorise a change of configuration when needed; data are no more dispersed in paper documents but embedded in a single PLM numerical platform with a unique management, and made available to each project stakeholder.

It also allows checking design completion before starting construction, and interactive numerical simulation of construction or maintenance sequences, to assess accessibility and optimise construction methods.

Major project stakeholders should be connected to the PLM numerical platform (civil works, conventional island, nuclear island, subcontractors) allowing a unified data management system, offering a novel design methodology and significantly reducing the project duration.

Digitalisation of the manufacturing chain, in particular replacement of paper documents, will also improve data traceability, components specifications and data sheets archiving and retrieving, and globally improve quality insurance, with a single source of data.

Intelligent design methodologies are of paramount importance to improve nuclear engineering, throughout the lifetime of nuclear projects, from cradle to grave.

Early in the process, at the conceptual design stage, key decisions structuring the project have to be made, including plant output, heat sink arrangement, general layout of the site, main features of the buildings (structure, large openings, etc.). All these aspects have to be jointly optimised, which is usually done in successive steps. 3D CAD tools should be used at the earliest stage possible in this design process, to accelerate the optimisation and assessment of alternative solutions ⋯ The seamless achievement of this step-by-step optimisation requires a high level of digitalisation, including static and dynamic calculations. As a lesson learned, it appears that simplicity and limpidity of the design, from any point of view, is the main objective of this conceptual design phase. Progress remains to be made in this field; software providers should consider the specific functionalities needed at these early stages of the design.

At the basic design stage, better optimisation and therefore improved competitiveness can be achieved by means of innovative numerical simulations. Qualified reliable simulation models, validated by experiments, can both be used to guide the optimisation process of designing, and to provide feedback to the design engineer. But they still require to be licensed by safety authorities, which sometimes proves to be highly demanding and may become a limiting factor of innovation.

Virtual/augmented reality technology aims to integrate the three-dimensional design models into a unified software environment, and capture all the geometrical characteristics of the different components into a full-scale digital 3D model. This allows a better optimisation of the reactor and auxiliary systems layout, and their interface with civil works. Improvement of industrial performance throughout the lifetime of a project, from design to decommissioning of a NPP can be achieved thanks to valorisation of data, as is being recognised in other industrial sectors: Predictive maintenance can be improved, using big data technologies to analyse and sort the huge volume of data collected by nuclear power plants.

Nuclear projects imply the management of a considerable amount of data, far greater than what is currently implemented in the automobile or aircraft industries. However similar PLM methodologies coupled with CAD tools prove to be very effective in improving data quality and streamlining project management. These processes will provide better traceability of the projects and benefit quality assurance and safety. Important gains are still accessible which need cooperation of all stakeholders including safety authorities.

Furthermore, digital simulation should be used to animate full-scale simulators for training purposes, and to improve emergency preparedness and severe accident response capacity.

For more than twenty-five years, instrumentation and control systems of nuclear power plants have made extensive use of digital technologies, However, many challenges are still lying ahead in the application of these technologies to the nuclear industry, including the qualification of both hardware and software for meeting safety requirements. Taking science and technology to market is key for nuclear energy competitiveness, but requires involvement of industry and regulators. Work remains to be done to develop codes and standards applicable to key software, when they are safety relevant. Besides, there are no international regulations and laws in the field of prevention and detection of incidents like hostile cyber-attacks of NPPs. Thus, an effort should be made to strengthen the establishment of certain legislation, and simultaneously attention should be given to relevant safety protection techniques relying on physical separation, firewalls, and effective management measures.

Conclusions

The nuclear industry would greatly benefit from full digitalisation at all stages from conception through construction, operation and maintenance up to decommissioning of reactors. Product lifecycle management, which is a tool for doing just that, is widely and successfully used in the aerospace industry. Other tools include numerical simulation, 3D CAD, and virtual/augmented reality. As a result of using these tools, performance would go up substantially, construction delays and cost increases would be avoided and competitiveness would be augmented.

Use of simulation tools (neutronic, thermohydraulic, etc.) at the design stage, is quite promising. For many years, digital instrumentation and systems have been used in nuclear power plants. In these two areas, a coordinated approach by safety authorities worldwide is needed.

In any case, utmost attention must be given to cyber security in order to detect and prevent cyber-attacks in a timely manner.

Chapter 9   Importance of nuclear research facilities and infrastructures

Recommendations

Two recommendations can be formulated covering the missions of nuclear research organisations in established as well as in emerging nuclear energy countries:

- In a globalised world, and taking into account the increasing cost of nuclear research facilities and their relatively low utilisation rate (at least for some of them), it is recommended to explore the mutualisation of large and/or heavy and/or specific nuclear facilities: research reactors, hot labs, irradiated materials labs, simulation and computation centres. This is valid for existing facilities and even more so for new ones.

- To allow nuclear emerging countries to have access to R&D facilities at acceptable cost, it is recommended that the R&D organisations and equipment makers in well-established nuclear countries make available low-cost laboratory facilities to address specific as well as general needs for experimentation and research in the nuclear field, including front-end and back-end activities.

The nuclear infrastructure in nearly all industrialised countries has originated from National Atomic Energy Commissions, themselves strongly marked by an academic and research culture. In addition to their own nuclear research and education centres, these institutions have also been the ultimate source of most of the national nuclear safety authorities as well as of technical support organisations (TSOs) for safety, of waste management agencies, and of a part of the nuclear industry.

At an early stage, military applications, where they existed, have been separated from civil applications, both in terms of programmes, infrastructure and human resources so as to ensure confidentiality and non-proliferation. With this caveat, civil nuclear research remains-to a large extent-non-secretive, obeying IAEA transparency rules, and exposed to a globalised world.

Very early-on, a separation appeared between the nuclear reactor industry and the fuel cycle industry. The former quickly became engaged in a competitive market, with some of the main actors originating from their national energy commissions, while others, such as Westinghouse, emerging independently from these agencies, designing, proposing and building power plants, supported by the massive streams of cash earned from the operation of reactors. Westinghouse, a brilliant success in this field, has strongly influenced the nuclear power industry in France, Japan, Korea and China. In the Gen-Ⅱ and-Ⅲ market segments, the role of nuclear research centres has often been, but by no means always, limited to case-by-case support, especially for high-level unexplained problems encountered during the operation of reactors by vendors or utilities.

Nuclear reactor fuel performance continues to be investigated at some of these centres while research on reactor life extension is also carried out by national laboratories. In almost all cases there is a need for specific facilities located in the nuclear research organisation (research reactors, test facilities, etc.). In France, research facilities are shared between by EDF and CEA.

As can be concluded from the above, the fuel cycle industry remains more closely linked to nuclear research organisations. The companies acting downstream in that domain have often been formed as "spin offs" of such research organisations. They became standalone companies, even when the founding research organisations kept a strong capital share, developing tight cooperative research programmes, and using more of the common innovation capacities (reprocessing or waste management, etc.). In this field, a close connection was needed to cope with the requirements of nuclear materials control. On the other hand of the fuel cycle, the enrichment business is mostly in the hands of private companies doing their own research.

During the last fifteen years, several issues have had a strong impact on nuclear research organisations and, in the following context, their role or at least their way of operation had to be re-defined:

- Lifetime extension for the most advanced Gen-Ⅱ reactors;
- Arrival of the Gen-Ⅲ designs;
- Preparation of the Gen-Ⅳ designs;
- Access of emerging countries to nuclear power or their desire to access;
- Severe accidents;
- Global increase of nuclear costs, largely consequences of increasing safety requirements;
- "Phasing out of nuclear" by some large industrialised countries;
- A huge need of nuclear educated manpower to replace the "Gen-Ⅱ staff" in conjunction with a relative disaffection of young people for scientific and technical jobs, at least in western countries (refer to Chapter 10);
- Strong increase of the decommissioning programmes and the associated feed-back;
- And last but not least, the emergence of the SMRs paradigm.

In the absence of evident and declared private financing, a good solution to house and manage an innovative nuclear project (or programme) might be the nuclear research organisations, where they still exist. Their experienced teams are capable of coping with short-term studies-while leaving time to select or build industrial actors-as well as with mid-or long-term studies such as those related to the fuel cycle (Examples: Gen-Ⅳ, SMRs).

However, in many emerging countries, the R&D organisations are managed by universities, professors and academic staff. This provides an acceptable basis to start activities but it should rapidly be supplemented by more industrial or at least more technological profiles in order to bring technical expertise and added value to the industrial actors of the nuclear programme when it is launched.

This should not however preclude to take into account disruptive innovations which originate most frequently from universities and academic research. Building efficient partnerships between academic teams and industrial companies is always helpful.

Conclusions

An emerging nuclear country, even if it has chosen to make a comprehensive appeal to external competences (which is the case when facilities are contracted under a build own transfer (BOT), or build own operate (BOO) model), shall have a minimum of technical competences to manage the relationship with the utility and to fulfil its responsibilities to ensure the safety of its operations vis-à-vis its population and the worldwide nuclear community. This role of "national owner's engineer" especially for safety issues should ideally not be sub-contracted. However, the support of foreign TSOs can be helpful and the experience gained with them in other countries could be shared.

In this framework, the role of a NEPIO (Nuclear Energy Programme Implementing Organisation) as defined by IAEA could be complemented with R&D missions, in support to the construction programme implemented by the NEPIO or by another national organisation.

Chapter 10   Challenges for education and training

Recommendations

The Academies consider that it is mandatory to consolidate education to satisfy the need for adequately trained engineers and technicians and to reinforce the attractiveness of the nuclear industry through more intense interaction between science and technology, industry and universities, and the prospect of interesting careers. They recommend to:

● Improve teaching methodologies, integrate modern IT, such as simulation, multimedia, e-learning, and virtual reality in courses for engineers and technicians, but also diversify courses to include a wider range of skills and competences such as PLM as well as cultural heritage;

● Enhance cost-effectiveness and safety of nuclear power plants, integrate the use of digitalisation for construction, operation, and maintenance in student courses;

● Take full advantage from available experience and help providing better training services to third countries, start cooperation in this area between France and China.

The context for education and training has deeply changed during these last ten years.

(1) The engineers, scientists and technicians graduating between 2020 and 2030 will be working until between 2065 and 2075 respectively. Those working for the nuclear industry will have to:

● Operate, maintain and modernise the installed nuclear fleet, but also dismantle part of it;

● Build new "Gen-II, -II+, -III", plants and then operate and maintain them;

- Conceive new plants, "Gen-Ⅳ", SMR, ⋯;

- Develop advanced fuel cycles, including waste management.

In France, even though the percentage of nuclear engineers is now superior to what it was 20 years ago, the nuclear industry is still in need of technicians (Professional baccalaureate, baccalaureate+2 years professional training and education and baccalaureate+3 years professional training and education), making up around half of the employees.

China keeps constructing nuclear power plants for more than 30 years. Especially after 2005, when the construction of a new cohort of NPPs was started, more and more universities in China were setting up nuclear technology courses, trained qualified personnel in highly valued competences for the industry. However, at present, their professional skills are still insufficient—employers have to systematically train staff for long periods and provide on-the-job practising to get them to the stage where they are authorised to effectively take on responsibility.

(2) The development environment of the nuclear industry is very different from what it was in the 80s; the nuclear power industry is at a crossroad now. Firstly, the necessity of dealing with climate change has become a global consensus. Green energy and low-carbon transition policy provide further opportunities for the development of nuclear power in China and for French-Chinese cooperation in developing a nuclear power market in third countries.

Secondly, as the technology of wind and solar energy improves rapidly, their operational costs decrease rapidly, but with more stringent safety requirements for nuclear power after the Fukushima nuclear accident, the needed investments for new NPP projects and for safety improvements of the existing NPPs are increasing significantly. As a result, the relatively strong competitiveness of nuclear power is weakening.

Thirdly, as the public is more concerned about nuclear safety, the continued strong development of the nuclear industry faces challenges, such as how to communicate with stakeholders and win their support.

Fourthly, due to more unstable power capacity and deregulation of the electricity market, sustained revenue from NPPs becomes uncertain.

(3) Safety issues have become extremely important for the nuclear industry; conservatism is encouraged while innovation is discouraged. For example, the "digitalisation" of the nuclear industry is lagging behind the digitalisation of other industries such as the aviation industry. But with the technology development of the internet, big data, artificial intelligence and others, more digital and intelligent technology will be applied in such areas as data management, product life-cycle management (PLM) and digital simulators (DCS). The required knowledge in these areas has to be taught to many employees in the nuclear industry and to government regulators, now and in the future. It is a precondition for its sustainability.

(4) Due to the complexity of nuclear safety issues, all nuclear industry employees should have a wide range of professional knowledge, skills, experience and cultural heritage. So, we must establish adequate knowledge management and personnel training systems for new employees to provide full training, to have competent teachers, and to promote more efficient intergenerational institutionalised cooperation. This applies to all employees, at all levels.

The responsibility to deal with the four preceding factors is shared between the nuclear industries and the "educational ecosystems" (universities, engineering schools, research institutes and technological institutes) of our countries:

● For the educational ecosystem, it is the responsibility to refocus student's curricula to include PLM, environmental issues, interdisciplinary and systemic issues, modern instrumentation, etc. and not only, as in the past, fundamental sciences (e. g. neutronics, thermal-hydraulics, strengths of materials, structural mechanics, etc.) and/or basic engineering and classical project management.

● For the nuclear industry, it is the responsibility to recruit employees with more diverse backgrounds, thus taking into account new ideas and ideas from "outside", fostering actively and steadily innovative engineering approaches.

● To promote nuclear safety culture and comprehensive quality management, the cooperation in personnel training in the concerned nuclear industry sector should be strengthened, training resources should be shared, and training standards should also be unified.

- Between the two, it should be made possible that scientists and engineers from industry work part-time or for a few years in education and, symmetrically, that professors/researchers work a few years in the industry.

It should be mentioned here that China and France have a number of common international cooperation initiatives regarding training of nuclear industry personnel, such as the promising experience with the young IFCEN, the Sino-French Institute for Nuclear Energy, part of the Sun Yat Sen University of Guangzhou. IFCEN delivered the first 80 engineering degrees in June 2016 with an innovative curriculum implemented by a Sino-French team.

With the support of IAEA, the Chinese government continues to train nuclear professional masters and doctors for emerging countries in the Harbin Institute of Technology. In order to support the nuclear industry training programme for students from emerging countries at the Tsinghua University, the Chinese Ministry of Education provides scholarships to 30 international students every year. Customised training for the nuclear industry is popularised at Chinese universities; the training of talents for nuclear power applications is speeding up. Based on the win-win training cooperation between university and enterprise, graduates get employment opportunities and enterprises get trained personnel. Training cooperation increases graduates' education in skills, accelerates their formation according to responsibility standards, and improves the level of the personnel teaching courses.

Foremost, the nuclear industry must evaluate its needs, takinginto account the age pyramid of its current staff. This difficult task has already occurred once in France, 15 years ago, when the retirement of the "baby-boomers" post-World War-Ⅱ after around the year 2000 had to be anticipated. Nowadays, in France, we need to anticipate the wave of retirements linked to the large number of recruitments made during the short period of accelerated construction of the majority of the French nuclear fleet.

When training the current employees at the different professional levels of companies and for the different types of jobs, we have to take into account not only the evolution or even outright replacement of systems, the upcoming of new software-which has always been the case, but also an altogether different education of the workforce. For example, the mindset, the values, and the learning habits of the new generation of employees in the nuclear industry; all of this is changing. The increasing performances of the IT world allows the use of multimedia, videos, e-learning, virtual reality to educate the employees at all levels-not only the staff and the engineers! Therefore, nuclear industry regulators should adapt, promote and support new training techniques to be applied in the nuclear industry.

For example, simulators have been used for a long time for the training of plant operators. With the progress of modelling, simulation and computer-power, simulators are becoming more powerful, in particular for simulating post-accidental situations, which leads to better training possibilities for plant operation. Moreover, the progress in the area of simulation, of virtual reality, etc. opens possibilities for differently and more efficiently training maintenance operators working in the field, especially those who have to intervene in areas where radiation imposes short intervention times. Here again, what is done in other industries, may be applicable and of help for the various operators working in the nuclear industry. The training centre of CGN innovated in a variety of simulation training courses and tools, integrated learning, teaching, training and research progress, and formulated multi-level training systems and facilities. They have developed stand-alone versions and portable simulators; "door-to-door training" and "learning while practicing" is becoming a reality; these innovations enhance the quality of training.

To succeed, the nuclear industry has also to keep current talents and attract new ones. Indeed, the nuclear industry is no longer regarded as the most high-tech industry, nor is it seen as the most desired industry. Renewed interactions with sciences and technologies, through up-to-date training of present employees and education of future employees are key.

Those countries that desire to develop and use NPP technology should encourage the cooperation in personnel training for the nuclear industry, especially the identification and training of talents for international projects needed for promoting the development of nuclear power programmes at the international level. Furthermore, the development and application of international nuclear standards, of local regulations and technical standards, as well as of the technical knowledge and competence for NPP design, construction and operation should be accelerated. As China and France have built a good nuclear professional teaching system, have a good international professional training basis, both countries would like to cooperate in this area and provide better training services for third countries.

Conclusions

In both France and China, there is a need for better education and training of more nuclear engineers and technicians. Furthermore, in about 20 to 25 years' time, there will be a second wave of retirements in the nuclear industry in France, leading to the need for accelerated recruiting. This situation needs to be anticipated. Attracting new talents would be helped by renewed interaction with science and technology. Professors/ researchers would benefit from opportunities to work in industry whilst industrial scientists and engineers working part-time in education would enrich training programmes. Training programmes would also benefit from modern IT, allowing cost-effective training through simulation and other teaching methods. Furthermore, the training programmes should encompass a wider and more diverse range of skills and competences. While technologies for wind and solar energies improve rapidly, reducing their costs, the nuclear industry is suffering cost increases due to more stringent safety requirements. In order to counteract such cost increases, today's students need to be familiarised with the digitalisation of all phases and elements of nuclear power plant construction, operation, and maintenance.

Chapter 11　Engineering and managing nuclear projects

Recommendations

- Nuclear projects are complex. The Academies recommend using the most modern tools to alleviate this complexity.

- Risk assessment should be performed throughout project implementation.

- Because changes in the safety regulations during project implementation brings major engineering uncertainties, these regulations should be fully validated and frozen before starting a project.

- A proper use of up-to-date management tools allows to cope with the engineering challenges of such large projects.

The size and complexity of nuclear power plants is exceptional: not including sub-components, several hundred thousand objects have to be designed, manufactured, erected, tested and commissioned, which is usually one or two orders of magnitudes more than projects developed by the aerospace or automotive industry. In addition, complex safety requirements, and quality classifications are to be complied with. For these main reasons, project schedules span over five years or more, not including site preparation, procurement of long lead items, and development of all basic and most of the detailed design. Therefore, controlling nuclear projects is a challenge, and insufficient control may lead to dramatic consequences. As an example, after the TMI accident (1979) new safety requirement were developed; therefore, on-going constructions were stalled pending the enforcement of these requirements. Consequently, schedules of many projects were doubled, and many projects were cancelled. More recently, recent Western projects faced difficulties due to insufficient completeness of the design before the start of construction, and lack of recent experience throughout the supply chain. On the other hand, many Chinese and other Asian projects provide evidence that the inherent complexity of nuclear projects can be overcome, and that implementation of appropriate design and project management methodologies result in well controlled project implementations.

The risks are even greater for FOAKs (first of a kind reactors). For these projects, R&D and technical verifications at the design phase are the key of risk control. In general, active and effective measures to ensure that R&D results are available at the design stage are paramount. Therefore, the R&D of new technologies needs to be integrated within the project schedule. Furthermore, licensing of FOAK generally proves to take time, as the regulator requires evidence of the effectiveness of new features. Therefore, FOAK projects require far greater attention, and construction should not start before new concepts are suitably validated and licensed.

When a nuclear project is based on a reference plant, the main difficulties and risks faced during its implementation relate to the quality and stability of the design, the performance of the supply chain, the delivery of all components at site in time and according to schedule and the proper coordination of all site activities.

Experience gained with recent projects proves that at least 70% of the detailed design must be completed before starting construction of any nuclear building, which requires that procurement activities are well advanced in order to get all input data needed for static and dynamic analysis, definition of anchors, construction plates and openings, etc. Only the definition of small items with no influence on the civil works and piping and electrical layout may be left behind. Design work must be fully and seamlessly integrated between the architect engineer and the contractors and suppliers, using the same CAD tools and data bases (refer to Chapter 8). Stability of the licensing requirements is a prerequisite, and regulators should understand that "combined construction and operation" licenses (COLs) provide the frame for better safety and quality.

Procurement activities and the strength of the supply chain are another key factor of success of nuclear projects. China and France are used to the owner's own model. In this model, the owner has its own design and project management in-house capacities. This model proves successful, so long as the skills, methods and tools are kept up to date, making use of the best tools and experience available on the market.

Coordination of all activities during design, procurement, erection and commissioning requires sophisticated project management tools, combining a huge component data base, a scheduling tool, a documentation control system etc. Such coordinated tools (project life management) exist; stringent procedures have to be defined and enforced to fully benefit from their powerful capacity (refer to Chapter 8).

Finally, strong controls need to be put in place, encompassing quality, safety, scheduling, and cost. To the maximum extent possible, they need to be independent of project management itself, and to have their own reporting line to the highest level in the organisation.

In order to guarantee the quality and schedule of nuclear power projects, necessary measures need to be taken to reduce construction costs, while strengthening the cost control. It allows reducing the project cost andimproves cost efficiencies of the nuclear power project. Cost linkage management can benefit from earned value evaluation. Earned value management can predict progress and cost deviations and propose measures to ensure that project objectives are achieved.

Appendix C provides a more detailed presentation of project risks. Sound design and project management has the common result of better controlling and alleviating these risks.

Conclusions

The complexity of nuclear projects needs to place a maximum of attention in the design quality, the robustness of the supply chain, the control of the consistency and schedule of the project with use of modern CAD and PLM tools.

Recent experience shows that nuclear projects, if properly controlled, can be delivered in time and within budget, while coping with stringent safety and quality requirements.

Methodologies, often implemented in other industries, are available and should greatly benefit the nuclear industry.

Appendix C　Main risks of nuclear power project construction

1. Financing risk

Nuclear power projects belong to investment in the fixed assets, in which steady cash flow is the priority of success when executing. The proportion of debt capital to equity capital depends on the balance of the fund cost and acceptable risks. In general, project financing risk covers all the risks of the project. Specifically, the significant project financing risks are debt paying ability; investment ability, refinancing, and financial risk (e. g. interest risk, exchange rate risk, etc.).

The project financing risk might be seriously influenced by duration in nuclear power plant project. Delay does not only cost more, but also leads to a series of legal problems if the power plant is not ready for commercial operation when the first loan instalments are due.

2. Design risk

Generally speaking, design risk of proven technology is relatively low, but the possible design change and documentation and drawing delay is still a bigger challenge to the project if the design work is not organised well on the basis of integrated project schedule. For FOAK (first of a kind reactor), R&D and technical verification at the design phase are the key of risk control. The key of new technology risk control is to take active and effective measures to ensure the rational coherence of R&D, design improvement and construction. At the same time, the R&D of new technologies needs to be matched within the construction progress schedule.

3. Purchasing risk

Developing procurement schedule should consider manufacturing cycle, required time of interface information for design. For long lead equipment, experienced manufacturers are needed. For the high-risk equipment, redundant procurement is recommended (the additional equipment can be used in next projects or increased cost is acceptable). Equipment with new supplier, new techniques and technology constitute the main risk sources for delay of design interface and equipment delivery procurement management. Mock negotiation simulating execution process is an effective method of risk identification including technology, schedule, cost, quality etc. And then targeted measures can be taken. Demand of localisation is acquired during the development of nuclear power in many countries. The risk of equipment localisation should be well assessed.

4. Construction risk

Large NPPs projects bear multiple construction risks, such asdesign changes, late drawings and instructions, lack of resources (especially qualified workers), accidents, equipment commissioning, etc. They must be accurately assessed, and impacts of quality and safety requirements on construction schedules shall not be underestimated.

Chapter 12 Assuring safety while keeping costs and complexity under control

Recommendations

- *The Fundamental Safety Principles* set forth by the IAEA are paramount to the development of the nuclear industry. According to these principles, nuclear safety requirements shall consider the latest state of science and technology. However, IAEA remains fairly general in defining any adequate level of nuclear safety, and the question remains: "how safe is safe enough?" The Academies recommend that IAEA reach a clearer view on this complex issue.

- Nuclear safety, for a very long time, will be regulated at national levels. However, the French and Chinese Academies recommend that safety requirements be harmonised at a worldwide level. Such harmonisation which happened more than eighty years ago in the aerospace industry is a prerequisite to industrial standardisation. The first step in this direction would be to reach a consensus on safety targets.

- The Academies recommend a "Risk-Informed Approach", as implemented by large nuclear countries, where safety requirements are balanced versus their benefits. The Academies question any approach requiring that "the best technology which is available" be systematically implemented, independently of its merits.

- While the Academies acknowledge that operating reactors should be regularly back fitted to always be "state of the art", they recommend that safety requirements be frozen during plant construction, and until commercial operation. Nuclear safety has to be seen in a comprehensive way therefore any change with even improved technologies may not necessarily have an overall positive impact.

From the onset of nuclear energy development, safety has been of overwhelming importance. The main concepts of nuclear safety (ultimate responsibility of the operators, independent safety authority, justification of facilities and optimisation of protection, prevention of accident by means of defence in-depth, mitigation by means of multiple and independent barriers, emergency preparedness and response) have been established at the early days of the industry. They are enshrined in the fundamental safety principles issued by the IAEA, and lastly revised in 2006.

Therefore, there is a broad consensus that proper nuclear safety requirements are the precondition for nuclear power development. It is also acknowledged that these requirements have to evolve overtime as a result of operation feedback, progress in science and technology and social demand for increased protection. Therefore, safety is a developing concept, and a relative issue. However systematically pursuing higher safety targets whatever the additional complexities that they imply is questionable.

Nuclear risk or safety levels should be considered in comparison with the other societal risks and be contained within a reasonable range. It is not appropriate to pursue higher safety blindly. While there was to a large extent, a common view worldwide on safety requirements of Gen-II reactors, inspired by the prescriptive objectives issued by the US-NRC, the development of Gen-III reactors reopens two main questions: ① "How safe is safe enough?", or "is there an appropriate safety level, to which all countries could agree? And ②Is it possible to develop a single set of safety requirements acceptable to all countries? It is easy to understand that a positive answer to question ①is a prerequisite to question ②. When comparing recent evolutions of safety requirements in different countries, it appears that regulators are not converging on these major issues.

Acknowledging that absolute safety cannot exist, and that a "residual risk" will always remain in the nuclear industry as in any other industry, it is necessary to have a clear methodology to decide upon the acceptable safety level.

This is answered by the IAEA safety objectives and principles, with the optimisation principle under which "The safety measures that are applied to facilities and activities that give rise to radiation risks are considered optimised if they provide the highest level of safety that can reasonably be achieved throughout the lifetime of the facility or activity, without unduly limiting its utilisation". However, if it is assumed that a zero risk should be pursued, or that there is never enough safety, optimisation means maximisation, which could lead to an endless process.

For sure, the nuclear industry must consider state of the art science and technology, but also ensure a better design optimisation, and not systematically add features and increase complexity. To introduce more rationality in the process, the "Risk-Informed Approach" should be followed, as developed initially by the US-NRC, and formalised by the IAEA. Under this approach, new requirements are systematically analysed inter alia by means of probabilistic assessments, to make sure that they add safety.

Best estimate analysis when considering beyond design basis accidents is more realistic and avoids over conservative approaches. More realistic design methodologies relying on a good understanding of the physical phenomena should be accepted, as overdesign does not only induce direct costs but may also lead to delays and difficulties in implementation.

Conclusions

Nuclear energy should not be looked at through the glasses of a risk analysis only. It should be considered using both a risk-and benefit-analysis and then weighing the risks against the benefits.

Nuclear safety is an ever-improving process; its progress has been impressive. After many years of development and integration of return of experience, the nuclear industry has reached a high level of safety and, in addition, nuclear energy is also low-carbon.

In order to improve competitiveness of the nuclear industry there is a need to implement methodologies relying on physical understanding and on qualified simulation models validated experimentally to avoid overconservative solutions. Interactions on technical matters with safety regulators are required to get these methodologies licensed.

It would be good and timely to pursue and deepen the harmonisation of safety regulations on a worldwide basis.

Chapter 13   Pertinence of international approaches to supporting the preparation of projects in emerging countries

Recommendations

To better support emerging countries nuclear energy programmes by vendor countries, progress should be made in two directions:

- Standardisation and stabilisation of the licensing (and regulatory) processes;
- Generalisation of long-term contracts for electricity purchasing guaranteed by the concerned governments.

Most emerging countries have an urgent need for stable and reliable base load electricity, to develop their infrastructures, their industry, and to cover the needs resulting from the accelerating growth of large cities. The nuclear energy option offers a potential answer to these requirements, supplemented by renewable energy sources, which may appropriately cover part of the needs of the population at the local scale.

At present, there are more than 50 countries positively considering developing nuclear power, most of which are emerging countries, that lack experience in nuclear power construction and feature an infrastructure in nuclear power that is underdeveloped. Therefore, IAEA has established a guide for the construction of nuclear power infrastructure in emerging countries (NG-G. 3. 1 document *Milestones in the Development of a National Infrastructure for Nuclear Power*). This guide points out three phases that emerging countries usually experience when they introduce the first nuclear power plant, and 19 infrastructure issues that need to be considered more specifically. Combined with the international experience, each emerging country can use this guide to choose its nuclear power development pattern which best suits the domestic specificities.

Furthermore, the philosophy of the international nuclear community, throughout IAEA, is that every country may have access to civil nuclear energy while respecting its proper timeline and situation to fulfil the conditions for maintaining, first of all, the high level of nuclear safety reached by the international nuclear community. The interdependence of all nuclear countries with respect to the prevention of accidents is undoubtedly a constraint. It shall also be an asset when the "old" nuclear countries help the "new" ones to overcome the difficulties and develop their projects. The cases of China and UAE illustrate two different but pertinent tracks.

China has chosen a nuclear power development pattern of introduction, absorption, innovation and substantial planning corresponding to its very large internal market. During the self-reliant design and construction of the Qinshan Nuclear Power Plant in the 1980s, China introduced the nuclear island technology from France and the conventional island technology from UK, and eventually built the Daya Bay Nuclear Power Plant. During the next 20 years, China has gradually formed its own technology system by means of introduction, absorption and innovation and has achieved standardised and mass construction. On this basis, China has introduced third generation nuclear power technology from the USA, France and Russia, from which the advanced design concept was adopted. Combined with proven experience in the domestic construction and operation of nuclear power plants, China has formed its third-generation nuclear power technology-HPR1000 with fully independent intellectual property rights. In this process, China has made great progress in its nuclear power infrastructure, and has gained the ability of a complete nuclear industry exporter.

UAE has chosen a nuclear power development pattern of comprehensive introduction. UAE has adequate reserves of capital and a concrete planning of nuclear power. To rapidly build up the nuclear power system, UAE has chosen a comprehensive introduction pattern corresponding to its lack of experience in nuclear energy research and relevant technology resources. The first nuclear power plant in UAE relies on the Korean nuclear power technology. The design and construction of the nuclear power plant was outsourced to South Korea, and the nuclear fuel supply was outsourced to France and Russia. With regard to human resources, the UAE nuclear energy agency hires experienced talents directly from overseas to take charge of the supervision of nuclear in UAE. Furthermore, UAE has established the International Advisory Board (IAB) formed by renowned nuclear experts from different countries, so as to promote the development of its nuclear power infrastructure.

More generally a nuclear newcomer country faces difficulties at three levels:

- Management

- Industrial

- Financial

The first management difficulty, results from the lack of nuclear competences in the emerging country, even if it is-and this need is all too often underestimated-only to play an ownership role. These owner-competences have either to be built mainly on a national basis. This has been the case for China, or to be in-sourced by making appeal to foreign competences or experts, like for example the United Arab Emirates (UAE).

The second management difficulty is the lack of familiarity with the specificities of nuclear projects; but this can easily be overcome with the help of IAEA or nuclear advisory companies.

The third management difficulty is due to the fact that a nuclear project is a programme involving the whole country, which is generally not the case with a classical engineering project. This is perfectly described in the IAEA NG guide.

However, among the 19 items of the milestones process, and with feedback from recent programmes, three issues appear to be of utmost importance:

- A safety and regulatory national system

- A robust financing scheme

- A long-term and reliable strategy to cope with spent fuel and other nuclear waste

One can easily understand from these considerations that the ability and the long-term commitment of the government of the emerging country are mandatory.

The industrial difficulties are mainly related to the fact that an emerging country has in general a relatively weak industry or is at least not familiar with the nuclear specificities and the very high level of requirements in terms of quality, traceability, auditing and control processes. On the other hand, the government generally wishes to reach as rapidly as possible a significant local share of the total project work.

Financing difficulties arise more and more often when the preparatory work of the NPP project arrives at the crucial stage of building the owner company with its shareholders, its governance, its financing means and the ways for return on investment to the shareholders. These items are related to one or several of the following issues:

● The high level of investment required (several G€)

● The limited capacities of the host country

● The constraints burdening the potential lenders as a consequence of Basel I, II and III accords regulating the banking industry

● The reluctance of the host country governments to guarantee a purchasing price of electricity at rewarding conditions for the owner company on a long-term basis

Only international approaches seem pertinent to support the emerging country decision-makers and overcome all these difficulties.

Regarding the management difficulties, bilateral governmental agreements and solutions by twinning appear to be the most adapted for appropriately putting in place a safety authority and a regulatory framework, for organising education and training to build up a human capacity, and for mobilising-on a long-term basis-a capacity of technical support such as research reactors, hot labs or waste labs. Site selection studies could also be added to this list, although the vendor consortium more and more frequently does the characterisation of the site.

The building-up of a first stage of national industrial companies, capable of taking in charge-with the required level of quality-specific work-lots of nuclear projects, is often a shared mission between ministries and industry associations or trade and industry chambers. The most efficient approach seems to be the creation of joint undertakings between the pre-selected national companies and foreign companies that are already experienced with nuclear contractors, or at least setting up a twinning consortium with these companies.

Financing difficulties have until now been overcome mainly by an important involvement of the vendor country(s) and its industrial companies:

● Through equity by the nuclear utility leader of the future operations, and sometimes of the reactor vendor

● Through debt with lenders or through export credit agencies

The use of build own operate (BOO) and build own transfer (BOT) contractual schemes, such as those employed in Turkey, is a real novelty in the nuclear domain, although it is currently used in other types of power production schemes.

As a counterpart, long-term contracts for electricity purchasing are mandatory to assure the return on investment. But this system is reaching some limits. Even Russia, who was the most important player in that game, is now facing financial constraints. Moreover, in most countries an excessive confidence in pricing of electricity by the free market reduces the ability of governments to establish long-term contracts, despite the remarkable example given by UK with its feed-in tariffs.

The standardisation and stabilisation of the licensing processes all over the world would give confidence to investors in the NPP projects scheduling.

Generalisation of long-term contracts for electricity purchasing guaranteed by the government would also enhance investors' confidence in the reliability of the forecasted return of investment. Both of these improvements rely on an international cooperation for which a forum to establish the above does not yet exist.

Lastly small modular reactors, developed by all the nuclear "vendor" countries could represent a complementary solution in a near future, as they can significantly lower the cost of the initial investment, and simplify and accelerate the implementation. They would probably require an increased level of cooperation between the vendor country and the emerging country especially in terms of regulations, waste, and insurance.

Conclusions

When developing nuclear power, different countries should choose the development pattern which best suits their domestic conditions, and gradually improve their ability in nuclear power development through international cooperation. International approaches and bilateral cooperation turn to be unavoidable for supporting emerging countries in introducing civil nuclear power in their energy mixes. Only they can help solving the management, industrial and financing difficulties arising on the way.

Out of consideration of the security of global nuclear industry as a whole, countries that have gained the ability of nuclear industrial chain supply, such as China and France, should take the initiative to serve the global community by sharing their nuclear energy resources and helping emerging countries to develop nuclear power in a safe and effective way.

Chapter 14　Human health assessment over fifty years of nuclear power activities

Recommendations

- Fifty years of normal operation of commercial nuclear reactors demonstrate that their radiological impact is extremely low, and well below the level of natural radiation. This fact should be better communicated to the public. It would also be important to report on the health effects resulting from fossil fuels like coal combustion to provide a proper perspective.

- Lethal radiological consequences from severe accidents like Chernobyl and Fukushima were limited; however large territories had to be evacuated for long periods of time. It is therefore important to retrofit Gen-Ⅱ reactors to further improve the prevention of such accidents, and mitigate their consequences so that no countermeasures would be required except in the close vicinity of the plants and for a limited period of time. In that respect, the safety level of Gen-Ⅱ reactors should be improved to bring that level as close as possible to that implemented in Gen-Ⅲ reactors, as was done in France and China.

Since the nineteen fifties, the impact of nuclear electricity production on the health of operators and the public living around NPPs (nuclear power plants) has been the subject of numerous studies. Radiological impacts of exposures during normal operation of the nuclear fuel cycle, as well as in case of accidents have been thoroughly estimated. The United Nations Scientific Committee on the Effects of Atomic Radiation (UNSCEAR) scientific reports, among others, periodically summarise results of these investigations. The concerns about exposure to radiation are covered below. A short explanation about the dose levels and their health effects of radiation exposure is provided in Appendix D.

14. 1   Radiation exposure from normal operations

The annual dose due to natural radiation is around a few mSv/a. The impact of NPPs for individuals living in the vicinity is extremely small, between 1‰ and 1% of the natural radiation and is well below accepted ICRP standard which limit additional exposure to 1 mSv/a. The average annual dose to the workers of the worldwide nuclear fuel cycle (from uranium mining through fuel fabrication and reactor operation to reprocessing) in the period 2000—2002 reported by UNSCEAR is about 1. 2 mSv, while the occupational exposure limit defined by ICRP for workers is 20 mSv/a.

Regarding uranium mining for example in Canada, the second largest uranium producer worldwide, no increase in radon beyond normal background levels can be observed in the vicinity of uranium mines and the average dose for workers at uranium mines and mills is about 1 mSv/a, significantly below the regulatory limit of 50 mSv/a in Canada.

Indeed, any type of electricity generation, may increase the radiation exposure of the public and workers from activities in the life cycle from mining, construction, operation and decommissioning. An important exposure pathway is the discharge of natural radionuclides (radon and its progeny) from the soil and from geological formations. For example, to obtain the materials for construction of an electricity generation plant of any type, mining activities are needed especially for collecting metallic ores. The UNSCEAR 2016 report contains a comparison of radiation exposures from the different types of electricity generation and their upstream and downstream activities.

It is estimated that the coal cycle contributed more than half to the total collective dose (individual doses times the people exposed). The collective dose is the consequence of all discharges due to a single year's global electricity generation. That estimate was based on the assumption that the discharges are those of modern coal plants. The nuclear fuel cycle including electricity generation, on the other hand, contributed less than one fifth of this collective dose. It is counter intuitive to note that under normal operations, the coal cycle gives rise to a higher collective dose per unit of electricity generated than the nuclear cycle, and a significantly higher dose per unit of electricity produced than the other technologies evaluated, with the exception of geothermal power.

14. 2   Casualties and health effects evaluation of accidents at NPPs

Since the beginnings of nuclear energy to produce electricity, the world has experienced a few significant accidents, involving NPPs entailing the fusion of the nuclear core with large release-consequences.

The Three Mile Island (TMI) accident in March 1979 led to a partial meltdown of the core and resulted in the release of radioactive gases including radioactive iodine (0. 55 PBq) into the environment. According to the American Nuclear Society, using the official radioactivity emission figures, "The average radiation dose to people living within ten miles of the plant was 0. 08 mSv, and no more than 1 mSv to any single individual. " A variety of epidemiological studies have concluded that the accident has had no observable long-term health effects.

The Chernobyl NPP accident in April 1986 led to a total core meltdown and resulted in the release of radioactive gases (1760 PBq of radioactive iodine, 85 PBq of radioactive caesium) and materials into the environment.

One hundred and thirty-four emergency workers suffered acute radiation syndrome, of which 28 died from radiation. Among the recovery operation workers exposed with moderate doses, there are some evidences of a detectable increase in the risk of leukaemia and cataract. The occurrence of thyroid cancer among those exposed during childhood or adolescence has significantly increased due to the drinking of milk contaminated with radioiodine during the early stage of the accident. The thyroid cancer is a rare disease and its prognosis has been improved to a great extent. During the period 1991—2005, more than 6000 cases were reported around Chernobyl in contaminated areas, of these, 15 cases had proven fatal.

Apart from the thyroid cancer from exposure in childhood, the excess occurrence of any other solid cancer and leukaemia in residents of contaminated areas has not been observed. The UNSCEAR committee indicated that "most area residents were exposed to low levels of radiation comparable to, or a few times higher than, the annual natural background radiation levels", "not likely to lead to substantial health effects in the general population" but that "the severe disruption caused by the accident has resulted in major social and economic impact and great distress for the affected populations".

The Fukushima NPP accident in March 2011 was the result of an earthquake followed by a tsunami. This led also to the melting of three cores of the 6 NPPs installed on the shore of the Fukushima site and the release of radioactive gases (less than 500 PBq of radioactive iodine, less than 20 PBq of radioactive caesium) and materials into the environment.

No acute health effects (including deaths) was encountered directly from radiation. During the first year after accident, the average doses for workers involved in the mitigation of the accident and adults living around areas were about 12 mSv and 1-10 mSv, respectively, and about twice as much for infants. The doses incurred over the first 10 years are estimated to be twice those induced during the first year. The UNSCEAR report (2013) and white paper (2015, 2016) have reviewed the relevant studies since the accident. The committee considered "a theoretical possibility that the risk of thyroid cancer among the group of children most exposed to radiation could increase. However, thyroid cancer is a rare disease among young children so that statistically no observable effects in this group are expected. "

The UNSCEAR Committee also "noted that the most important health effects that had been observed among the general public and among workers were considered to be on mental health and social well-being. "

That statement would be valid each time large releases of radioactivity would occur. Therefore, a priority was to back-fit Gen-II reactors in order not only to reduce the probability of core melt even further, but also to allow a controlled release of fission products in case of containment overpressure, by means of filtered containment venting systems and other engineering features. The safety level of these reactors, after back-fitting is now as close as possible to the safety level of Gen-III reactors. If such systems had been considered in Japan, no long-term evacuations in and around Fukushima plants would have been needed. Such systems are implemented in China and France.

Conclusions

Over the last 50 years of NPPs operations, the feedback from experience of severe accidents show that they do not have any or only a very limited radiation impact on humans if fission products are contained. On the other hand, in case of dissemination, large territories may have to be evacuated. Health consequences to the public from NPPs accidents are also limited: health effects from ionizing radiation depend on the doses received and as shown above, such doses have been low. But the social consequences in terms of mental health and well-being at the local level near the accident sites have been important. This explains why considerable efforts are made to enhance safety in all current and future reactors with the objective of preventing release of radioactivity in any circumstances. Under normal operations of nuclear electricity activities, the levels of radiation exposure to public are very low and even lower than those induced for example by coal-fired electricity (coal naturally contains traces of radioactive elements such as U, Th, Rd, etc. , which escape into the atmosphere when coal is burnt in coal-fired electric power plants).

Appendix D   Dose levels and health effects of radiation exposure

People use and are exposed to non-ionizing radiation sources every day. This form of radiation is not sufficiently energetic to ionize atoms or molecules. Examples for such non-ionizing radiations are the electromagnetic fields in the neighbourhood of microwave ovens, radio sets or mobile phones. Some types of radiation, generically designated as ionizing radiation have enough energy to knock electrons out of atoms, upsetting the electron/proton balance of matter and leaving positive ions.

Humans are continuously exposed to ionizing radiation from natural sources ever present in the environment and artificial sources due to radiation application. The global average annual dose to an individual is estimated about 3 mSv (Table 2). About 20 percent of the exposure is from artificial sources, mainly from medical applications, for example, the dose from one computed tomography (CT) scan of the abdomen is about 10 mSv.

Table 2   Global annual doses of average public exposure by radiation sources
(*Radiation Effects and Sources*, UNEP, 2016)

Total natural sources	2. 42 mSv	Total artificial sources	0. 65 mSv
Food	0. 29 mSv	Nuclear power plants	0. 0002 mSv
Cosmic rays	0. 39 mSv	Chernobyl accident	0. 002 mSv
Soil	0. 48 mSv	Weapon fallout	0. 005 mSv
Radon	1. 26 mSv	Nuclear medicine & Radiology	0. 65 mSv

The energy of ionizing radiation can damage living tissue by killing cells or modifying cells. These health effects are dependent on the dose of radiation exposure. In UNSCEAR reports, the dose is described as high when it exceeds 1000 mSv, it is defined as moderate when it is between 100 mSv and 1000 mSv and it is low when it is less than 100 mSv.

If the number of cells killed by radiation exposure is large enough, it may result in tissue reactions and even death, e. g. loss of hair, skin burns and acute radiation syndrome. The severity of these effects increases with the dose when a certain threshold is exceeded. For example, a dose of more than 1000 mSv could cause acute radiation syndrome. The dose thresholds of possible harmful human tissue reactions are above 100 mSv. Such effects are observed in radiation accidents or in radiotherapy.

If the modification of cells irradiated is not repaired, it may result in cancer or heritable disease affecting offspring. These effects are stochastic, and the probability of their occurrence depends on the radiation dose received. The UNSCEAR 2010 Report indicated that "There is strong epidemiological evidence that exposure of humans to radiation at moderate and high levels can lead to excess incidence of solid tumours in many body organs and of leukaemia." For example, 10423 survivors of the atomic Hiroshima and Nagasaki bombings died from cancers (solid tumour in addition to leukaemia) up to December 2000, of which 572 cases (about 5%) could be attributed to the radiation exposure from the bombing. However, there is no clear evidence of excess heritable effects of radiation exposure in humans.

Chapter 15　Risk perception relative to real hazard

Recommendations

● More efforts should be made to promote and implement the nuclear safety culture, and improve a two-way interactive communication to help the public viewing the nuclear risks in rational and objective ways.

● More efforts should be directed towards explaining in simple terms what measures have been taken after Fukushima to strengthen the safety of the existing Gen-Ⅱ NPPs and improve the designs of the Gen-Ⅲ and -Ⅲ＋NPPs. With respect to the severe nuclear accidents, it should be especially emphasised: firstly the implementation of prevention features which reduce by an order of magnitude the probability of occurrence of such accidents, and secondly the limitation of any radiological consequences of such accidents to the close vicinity of the plant, and for a limited time period only.

Every time that a new technology emerges, two logics, almost metaphysically confront each other. One logic is reduced to the evaluation of the costs/benefits balance, the logic of operating and industrial players, who must innovate to remain competitive. The other logic pays attention to the negative consequences of the technological innovation and attempts to re-build an approach whereby "rationality" would impose limits to the conclusions from such costs/benefits estimations by taking into account other, more ethical, qualitative or indirect considerations.

Following this line of thought engineers and scientists are enjoined to focus all their efforts at avoiding-at any price-not only the catastrophe but also the shadow of a possible catastrophe! From evidence in most countries in the world, one can say that nuclear activities are supporting this goal and are therefore an obvious illustration of this observation.

Hence, it is necessary to recall that if the existence of a risk requires the existence of a danger, risk should not be confused with danger: risk is the multiplication of the exposure by the danger. If there is no exposure to a danger, there is no risk.

One can imagine that risk perception corresponds to the correct assessment of the risks themselves as soon as they are objectively proven. But this is a utopian dream, which can be analysed from three perspectives.

(1) Psychological. Risk perception does not only rely on non-rational factors. Everybody builds his/her own "basket of risks" and manages it with a certain level of rationality. For example, some risks are accepted according to the compensation they provide; this is for instance the case for the inhabitants around a nuclear power plant or facility.

(2) Sociological. The perception of risks strongly depends on culture, country and history. Moreover, risks selected by an individual are not dreaded in the same way than risks experienced collectively. Nuclear energy is a mode of production of electricity that has not been explicitly chosen by most of those who could be affected by its drawbacks.

(3) Communication. The subjects of communication affect risk perception and risk value judgment of the public. Since nuclear energy is highly scientific and technical and most members of the public do not have much experiential knowledge on it, one expects that the media will perform a strong heuristic role in shaping public perceptions on nuclear risk a role that has become more pervasive in the current era of Web based media.

15. 1 Analysis of risk perception regarding nuclear energy from the perspective of psychology

Psychology surmises that the individual perceives risks through cognitive psychological mechanisms. The risk-related cognitive deviation is linked to the individual's cognition and judgment. The causes of risk-related cognitive deviation include: subjective factors such as individual personality traits, knowledge and experience, "loss" or "expectation of risk", and objective factors such as the nature of risk, size, degree of control and comprehension factors. Hence, for different kinds of people there are differences in their understanding of risks related to nuclear energy. For professionals, identifying the risk is usually based on technical evaluations weighing the pros and cons; on the other hand, the public tends to perceive its own interests closely related to safety and health, ignoring the benefits of nuclear energy, giving rise to the NIMBY[1] syndrome directed at nuclear facilities.

The public usually regards potential catastrophe, uncontrollable and unknown as the intuitive characteristic of nuclear energy, which was aggravated by the TMI, Chernobyl and Fukushima accidents. The public is not usually aware that the radiation effects of normal operation of nuclear facilities are much lower than those of the coal industry and that new Gen-Ⅲ and Gen-Ⅳ reactors are designed to prevent release of radioactivity in any circumstances. However, the public still believes that nuclear risk is much higher than that induced by other industries.

Results of a survey conducted by government departments, researchers, and the media of public acceptance of nuclear energy in China indicate that:

● The public disposes of limited information sources on nuclear energy and is in lack of basic knowledge.

---

[1]　Not in my backyard.

- A majority of the public worries about the development of nuclear energy is mainly caused by low public participation, lack of information transparency and apprehension about nuclear safety. As a result, only 40% of the public supports the development of nuclear power in China.

- The Fukushima nuclear accident has had the consequence that the public has become more sensitive to the possible development of nuclear energy projects, and is opposing such projects, especially near their homes.

In France, the main findings of the *IRSN 2016 Barometer on Risks and Security Perception by the French Public* are:

- Terrorism has become the first worry.

- More than one French in two declares that they have more confidence in science than ten years ago and also that they trust experts and technical and scientific organisations when intervening in the nuclear domain.

- The majority is in favour of giving the public access to the results of expertise.

- Although 46% of the respondents think that all precautions are taken to assure a very high level of safety in the French NPPs, about 90% think that an accident in a nuclear power plant would have very serious consequences.

15.2 Analysis of risk perception regarding nuclear energy from the perspective of sociology

The perception of risk of an individual is not just based on individual psychological cognition, but is also associated with the social organisation or social system. The individual risk perception is affected by social position, community influence, and social background. Disaster events interact with psychology, society, system and cultural status, which can profoundly influence risk perception. This is known as the social chain effect of risk perception: when fear is more frequently mentioned in conversation or in the media, the public pays more and more attention to the related information, and finally "perceives" (imagines) a risk, which is an actual strong deviation from the factual risk. For example, after the Fukushima nuclear accident in Japan, the United States, France, Germany and some other countries, the consequences of this deviation manifested themselves in the phenomenon of buying iodine tablets. In Malaysia, the Philippines and Russia, the phenomenon appeared in the form of buying iodine tincture, in South Korea it was seaweed product snapping, and in China buying of iodized salt. More importantly there are examples where nuclear projects have been brought to a halt because of public protest, a situation which needs to be carefully investigated to find better ways of managing such societal crises.

15. 3 Analysis of risk perception regarding nuclear energy from the perspective of communication

The subject of risk-related communication involves national government agencies, scientific experts and scholars, the public, the media, non-governmental organisations and so on. Different subjects of communication will affect the different risk transmission effects, especially the public risk perception and risk value judgment. After the Chernobyl accident, the various reports from the government, the media and organisations on casualties caused by the accident differed in a notable manner, leading to a situation where the public was no longer able to make the correct judgment regarding the real impact of the accident.

The public is highly sensitive to the acquisition and trust of risk information. As a result of incomplete risk information, such as government avoidance measures or ambiguous attitudes towards risk handling, the public's fears of risk tend to increase leading to a loss of confidence in the information provided. Media coverage and qualification of the "hydrogen explosion" in the Fukushima nuclear accident as a "nuclear explosion", also tended to promote public panic.

China is dedicated to improving its national nuclear safety system, enhancing nuclear safety capabilities and boosting nuclear safety culture. In 2014, nuclear safety was formally incorporated into the Chinese national security system.

15. 4　The problems, challenges and difficulties of nuclear risk perception

Risks are both objective and influenced by media communication, individual perception and judgement, and by social, institutional and cultural factors. With the emergence of nuclear power plant construction and accidents in the world, and the changes and development of nuclear power policies in various countries, the public's cognition of nuclear power has also changed, indicating a high degree of attention to this matter, a limited degree of acceptability and cognition, and subjective and irrational characteristics of attitudes. How to improve public awareness, regulate the dissemination of information and reduce public panic: these are the challenges and difficulties of global nuclear energy development.

To have a better understanding of what could be done to improve public awareness and in the light of the above considerations, it is of utmost interest to recall two major discrepancies between real hazard and risk perception in nuclear activities.

● The real hazard is that accidents at Gen-Ⅲ + nuclear power plants will not entail major radiological consequences in case of severe accidents (and will only require limited countermeasures).

The subjective public perception of the Risk is, however, that in case of airplane crash or a severe accident damaging the reactor core, sizable quantities of radioactive products would still flow out of the containment building.

● The real hazard is that the noxiousness of low radioactive levels and doses is still an important subject for biology studies, without clear conclusions for the time being. But the return of experience on populations living in regions with high levels of natural radiation (Kerala, Britanny, etc.) show no perceptible long-term effects.

The subjective public perception of the risk is, however, that the first Becquerel originating from a nuclear power plant or facility is dangerous for human health. But if that Becquerel is linked with radiography at a dentist or at a hospital, it is not.

Conclusions

The policies of full transparency and information of the public about the results of scientific and technological studies dedicated to nuclear issues and projects seem to be fruitful. They strongly contribute to reducing the gap between real hazards and risks perception by the public. However, the public generally overestimates the real hazards of nuclear energy because of limited scientific and technical knowledge. Work remains to be done to make the public aware of the huge progress in safety due to the post-Chernobyl and post-Fukushima measures put in place on all the reactors around the world: by means of adequate prevention features, the estimated probability of accidents with core melt has been reduced by a factor of ten, both for the existing fleets in France and China, and for Gen-Ⅲ reactors. Also mitigating features have been back-fitted in existing reactors, or implemented in the design of new reactors which would drastically reduce radiological consequences of such accidents; thus, no countermeasures would be needed, except in the close vicinity of the plant, and for a limited time only.

Chapter 16   Improvement of public awareness and governance requirement

---

Recommendations

- Countries developing nuclear energy must share information and arguments to explain the driving advantages and weaknesses of nuclear energy and remedies to enhance public awareness. The role of nuclear energy as a stable and reliable source of dispatchable electricity free of greenhouse gas emissions needs to be emphasised.

- Nuclear operators and stakeholders should have a positive communication strategy on the successful operation of NPPs and be transparent on disclosing events.

- Beyond the shared arguments and messages, communication strategies should take into account the different development rates of those countries that massively introduce intermittent renewable energies. There is no reason that such massive introduction be conflicting with nuclear energy if renewables serve to replace fossil fuel plants and not to reduce the already carbon free nuclear electricity.

For both industrialised and emerging, Western and Asian countries and more generally on a worldwide scale, the driving advantages of nuclear energy are:

- No emission of greenhouse gases, supporting the fight against climate change;

- Large power operated as base load, assuring the stability of the network and allowing the development of industry and heavy infrastructure at the level of a country;

- A today abundant and secure availability of low-price electricity compared to other sources;

- No need for storing electricity, which strongly penalises intermittent energies.

On the other hand, the driving weaknesses are:

- The burden of high-level long-lived radioactive waste disposal, admittedly with relatively limited volumes but during hundreds of thousands of years;

- The effects of severe accidents on large areas and significant populations.

In all these countries, the basic efforts to improve public awareness will consist in providing accessible information on each of these items, with special attention to quantifying the issues, and discussing solutions.

What distinguishes countries and requires a differentiated treatment, is the trend in electricity growth. In China and more generally in rapidly developing economies, the fast increase in electricity demand allows a progressive and substantial introduction of intermittent renewable generating capacity that does not oppose the simultaneous development of a strong nuclear energy sector. The situation is quite different in Europe, where the perspectives of electricity consumption are relatively flat (but might be on the rise again if fossil fuel consumption were to be replaced by electricity). Therefore, while a major and rapid introduction of intermittent carbon-free renewable energy (solar and wind) requires large investments, the main consequence will be to reduce the demand for nuclear generation. However, this has very limited impact on CO_2 emissions, as nuclear energy is equally carbon-free. Furthermore, it has also very limited effects on the installed nuclear capacity, which has to be kept in place to provide the necessary back-up for the intermittency of renewable energies.

The fashionable belief that the development of intermittent energies "allows to take steps in imagining the world of the future" together with more or less philosophical ideas like those of the "theory of negative growth" compounds the difficulty in Europe requiring new arguments and further efforts to enhance less emotive public awareness.

Further arguments and details related to the five items mentioned previously can be found in other chapters of this report. In particular, issues related to the management of radioactive waste are considered in Chapter 6 where it is shown that disposal of the long-lived radwaste in deep geological repositories is feasible. Accidents involving nuclear power plants and other installations are reviewed in Chapter 14 together with the considerable improvements that were accomplished at current installations and included in the design of Gen-III systems. However, additional considerations are useful for a better understanding of the present situation in Europe.

Currently, and probably in the coming decades, climate change issues will not be sufficient to convince public and political bodies to welcome new NPPs without the strong governmental support that one finds for example in Finland and the United Kingdom. Such governmental support must be available in the short-and medium-term during the decision phase (5 to 10 years) and the construction phase (5 to 10 years) but also in the long-term e. g. during the operational lifetime of the plant (at least 60 years).

If the contribution of nuclear energy to the reduction of CO_2 emissions is known to the public (but more or less denied by several anti-nuclear NGOs!), its value is not fully appreciated. The economic benefits (once again at least in Europe) in terms of generation costs are being questioned in a context where more stringent safety requirements have led to rising production prices, while photovoltaic production costs have significantly decreased.

Today, the technological, industrial and economic culture of politicians, media and the public in general is relatively limited and short-term oriented. Industrial assets and economic considerations are all-too often underrated in the balance with political or even philosophical arguments in some major decision-making processes.

Fortunately, there are political decision-makers who take in consideration economic objectives and technological constraints. This is the case with many of the members of the French Parliament belonging to OPECST (the Parliamentary Office for the Evaluation of Scientific and Technical Choices). This office carries out well-informed investigations and delivers high quality reports that are often ignored by the pubic and by the media but that deserve to be taken into account when deciding about complex issues in the energy field.

But in almost all these countries, the major misunderstanding lies in the pace of the "energy transition". There is no reason requiring a substitution of nuclear electricity by intermittent renewables in less than 50 to 100 years if this intermittency then requires back-up through fossil fuel fired electricity generation! This is the central additional message that must be offered to the public.

Public Awareness and Governance Requirements are very much influenced by Communication Strategies of nuclear operators but also of anti-nuclear NGOs. Utmost attention must be given to preventing that messages of nuclear operators be completely turned upside-down with respect to their initial meaning. As an example, the EDF motto "Safety First" has been interpreted by many people as "The EDF nuclear power fleet is currently unsafe" and needs urgent modifications. This interpretation is especially associated with the enormous amount of the investments required by the so-called "Grand Carénage".

In order to start building new nuclear facilities on "green field" sites, strong governmental support is needed and long and well adapted procedures are mandatory. This is mainly not due to local opposition but to professional international opponents and-at least in France-to a complex network of environmental laws, such as the Law on water, the Law on coastlines, the Law on biodiversity. Even the extension of existing sites may become problematic. In China, some inland projects are still on hold since the Fukushima accident.

Conclusions

The benefits of nuclear energy and in particular its stable and massive production of electricity with low emissions of greenhouse gases are not sufficiently publicised. The need for base-load production of electricity is not well understood by the public and there is an insufficient appreciation of the remarkable services provided by the electrical network, its stability and its relatively low sensitivity to meteorological conditions. It is necessary to bring more reality into the debate about the energy mix. In particular it must be emphasised that the intermittency of renewable energy needs to be compensated for: when renewables do not produce electricity because there is no sun or wind, then back-up capacity must be brought on line, as electricity storage facilities are not available, and may not be available for many years. If the conventional back-up capacity is then based on combustion of fossil fuels, it will induce emissions of GHG. In addition, when the amount of renewable energy generating capacity exceeds a critical level, there will be severe stability issues concerning vast swathes of interconnected grids. This is due to the necessity for conventional generating capacity to operate in a highly intermittent fashion, which may not be technically or economically sustainable.

It is worth explaining these fundamental issues to the public and showing that nuclear energy is needed to safely and continuously produce electricity in large connected electricity networks responding to the growing demand of an urbanised society. Because positions about the energy mix are often influenced by emotional arguments, there is a need for enhancing the level of understanding in terms of technological constraints and fundamental economic objectives and to provide further information on safety issues and on progress made to deal with accidents in a safe manner. This might be achieved with better communication strategies and improved education in energy matters so that people understand the issues without falling victim to misleading arguments and developing an opinion based on a rational analysis of facts.

It is also important to take into account countries' different development rates, which in cases of slow economic development may create conflicts between the massive introduction of intermittent renewable energies and nuclear energy, while in cases of rapid economic development they may not.

Chapter 17   Organisation, methodologies and roles of the different stakeholders to improve public understanding

---

Recommendations

To improve public acceptance, it is recommended that actions be taken at three different levels.

- On the technical level, it is important to take full account of experience drawn from major accidents to operate current power plants in a safe and efficient manner.

- On the organisational level, central governments should formulate their energy strategy and strive to carry it out. A clear separation is needed between operators on one hand, and safety authorities and their safety support organisations on the other hand. All actions at the local level to positively establish an accessible dialog platform for the public will improve confidence.

- On the communication level, it is important to make these efforts known to the public at large: organise access to transparent, exact and structured information. In order to take due account of social acceptability, it is important to develop education on energy issues, essential technical and economic factors, environmental impact and risks, and improve public understanding of fundamental energy challenges.

Experience drawn from the major accidents of Three Mile Island, Chernobyl and Fukushima, has led to reconsider the risk factors that must be taken into account in order to be better prepared in the event of having to face a large release of radioactivity into the environment and have stressed the need for independently and transparently informing the public on nuclear issues. This requires different types of action.

The first type of action is essentially technical and consists in making constant progress in plant operation in terms of efficiency and safety and integrating the return from this experience into the current fleet of nuclear plants and into the engineering design of future power plants. In order to manage risk factors effectively, they need to be rationally analysed and prioritised. This has led to third-generation reactors that have considerably improved characteristics of resistance to accidents and to various aggressive acts.

The second type of action is organisational and involves three levels of stakeholders:

● Central governments and their functional departments of energy, education, communication, health, etc.

● All stakeholders of the nuclear industry, which includes supervision departments, operators, safety authorities, independent technical supporting organisations, industry associations, etc.

● Local authorities and the public in the areas where nuclear projects are located.

At each of these stakeholder levels, different responsibilities and potential for actions can be identified:

(1) Central governments should harmonise their national framework in order to formulate their national energy development strategy, including for the development of nuclear energy. Then they should strive to pursue its implementation, and coordinate their subordinate functional departments to commit to the national energy strategy.

(2) All stakeholders of the nuclear industry have to structure the nuclear sector in a rigorous manner clearly defining their respective roles so that they engage and practice a safety culture. This is defined by the IAEA guidelines as "the assembly of characteristics and attitudes in organisations and individuals which establishes that, as an overriding priority, protection and safety issues receive the attention warranted by their significance." To implement this culture, it is appropriate to spell out the roles of the different actors:

- The responsibility of safety relies primarily on the operators of the nuclear reactors.

- The operations are placed under the administrative authority of an independent public safety agency, which controls all the civil nuclear installations. In France, its independence is guaranteed by the fact that its members are nominated for a six years irrevocable mandate. This agency organises regular, mandatory inspections of nuclear sites and equipment, delivers the authorisation for operations and has the power of interrupting them under all circumstances when it considers that this is necessary.

- The administrative agency needs to rely on a strong technical expertise; it can develop it in-house, or be backed by one or several Technical Support Organisations (refer to Chapter 7).

(3) At the local level, different situations can be encountered with respect to the type of decentralisation of the country. For example, in France, the local representative of the Central government (the "Préfet") is the only interlocutor of the operator. Local authorities are committed to consult and inform the public in the areas where nuclear projects are located: projects related with nuclear energy are formulated through complete decision-making procedures. After projects are approved and construction needs to be prepared, local authorities should, together with operators, positively establish accessible dialogue platforms for the public as well as transparent information disclosure mechanism. In addition, local authorities should actively explore ways to seek integrated development of nuclear projects and local economic and social progress including support in the form of financial contributions from the nuclear project and local employment.

In some cases, in France, public consultation of citizens needs to be organised prior to decisions, in order to better understand the source of fears. This was done for instance for the planned deep repository of nuclear wastes project named CIGEO.

The third type of action is to make these efforts known to the public by enhancing its awareness and understanding.

However, the lack of a basic education on radioactivity and the invisibility of radiation, the association in many minds of nuclear energy with nuclear weapons, leave much space for fears. Even for the most obvious benefits of nuclear physics, such as the irradiation of tumors to cure cancers, the use of nuclear magnetic resonance in medical imaging, the understanding of the history of our planet through radio-isotopes, the adjective "nuclear" is banned from the vocabulary as too scary. It has become clear that the future of the nuclear industry will depend to a great extent on public understanding and acceptance of this technology. This is illustrated in Europe by countries such as Italy that now refuses any nuclear industry, or like Germany that has decided after Fukushima to close its nuclear power plants by 2022. It is thus important to explain the safety measures that have been included in the new designs by assuring the transparency of information, providing the necessary education, and responding to growing demand for information from the public at large. It is also timely to improve the way we inform the public and communicate on accidents and incidents, be they natural or man-made. In France, at the national level, a high-level committee for the transparency and information on nuclear safety (HCTISN) guarantees that exact and accessible information on the civil nuclear operations is available to the public.

Conclusions

Beyond the many actions taken to insure the safety of operation of nuclear power plants, it is important to spell out the roles of the different stakeholders and provide transparent, exact and structured information. This constitutes the best response to different negative attitudes towards the development of nuclear energy. It is also important to underline the benefits of nuclear energy as a safe, clean and effective electrical energy and as an asset of economic development.

Glossary

ABWR: advanced boiling water reactor

APWR: advanced pressurised water reactor

ASN: Autorité de Sûreté Nucléaire (France)

BOO: build own operate

BOT: build own transfer

BWR: boiling water reactor

CAD: computed aided design

CEA: Commissariat à l'Energie Atomique (French Atomic Energy Commission)

CEFR: China experimental fast reactor

CGN: China General Nuclear Power Corporation

CNNC: China National Nuclear Corporation

COP: Conference of Parties to the United Nations Framework on Climate Change

DOE: Department of Energy, US

EDF: Electricité de France (French Utility)

EPR: European pressurised water reactor

ESBWR: essentially simplified boiling water reactor

EUR: European utilities requirements

FNR: fast neutron reactors

FOAK: first of a kind

GCR: gas cooled reactor

Gen-Ⅱ, Gen-Ⅲ, Gen-Ⅳ: refer to the second, third and fourth generations of nuclear reactors presently operated or under development. The first generation were prototypes, which are now decommissioned

GFR: gas cooled fast reactor

GHG: greenhouse gas

GIF: Gen-Ⅳ International Forum

HL-LLW: high level-long lived waste

HLW: high level waste

HPR1000: advanced pressurized water reactor developed in China (also named Hualong One)

IAEA: International Atomic Energy Agency

ICRP: International Commission on Radio Protection

INSAG: International Nuclear Safety Advisory Group (advising the IAEA general director)

IRSN: Institut de Radioprotection et Sûreté Nucléaire (France)

IT: information technology

LLW: long-lived waste

LLW: low level waste

LWR: light water reactor

MOX: mixed uranium-plutonium oxide

MSR: molten salt reactor

NEPIO: Nuclear Energy Programme Implementing Organisation

NGO: Non-Governmental Organisation

NNSA: National Nuclear Safety Administration (China)

NPP: nuclear power plant

NRC: Nuclear Regulatory Commission (USA)

PLM: product lifecycle management

PPA: power purchase agreement

PWR: pressurised water reactor

R&D: research and development

RW: radioactive waste

SBO: station blackout

SFR: sodium-cooled fast reactor

SG: steam generator

SMR: small modular reactor

SWU: separation working unit

TMI: Three Miles Island

TNR: thermal neutron reactor

TSO: technical support organisation

UNSCEAR: United Nations Scientific Committee on the Effects of Atomic Radiation

URD: utility requirement document (developed in the United States)

VHTR: very high temperature reactor

Postscript

In the present situation where much of the electrical energy production is based on fossil fuels and more specifically on coal, nuclear energy constitutes one of the most realistic options for supplying electricity in a safe, efficient and clean way and for simultaneously solving environmental and climate change problems. Because it is a stable and massive source of energy, it can reliably provide dispatchable electricity and can complement renewable electrical energy sources (like wind or solar) that are mostly intermittent and are not easy to mobilise to respond to demand.

The development of nuclear energy still raises many challenges and issues with regard to safety, management of long lived radioactive waste, development and deployment of advanced nuclear reactors, economics, public acceptance, etc. As two of the main countries with large capacities of nuclear power plants (NPPs), both China and France attach great importance to the peaceful use of nuclear energy in the world, and have the responsibility and willingness to help emerging countries in their development of NPPs and in their wish to resolve the challenges they will face.

In the continuation of COP21 and COP22 aimed at significant worldwide reduction of greenhouse gas emissions, the three Academies (Chinese Academy of Engineering, National Academy of Technologies of France and French Academy of Sciences) believe that their initiative to shed light on some of the complex issues related to nuclear electricity generation could send a strong and valuable message to other countries' academies, decision makers and society in general.

The present report reflects positions of the three Academies acting as independent bodies, and shall not be construed as positions of industrial actors in the NPPs field or positions of either the French or Chinese governments. In this report, the contributing academies aim at outlining the history and perspectives of nuclear energy, and address the key issues to be considered in order to make nuclear energy even safer and affordable for the benefit of developed and emerging countries. Although this report touches on many issues, it is not intended to be exhaustive. It synthesises reflections and discussions carried out over a period of six months.

The report comprises a synthesis and a set of seventeen chapters. Two chapters give a brief account of the history, problems and challenges of nuclear development and address issues concerning the deployment of Gen-III NPPs. Two chapters deal with scientific aspects, including promises and challenges of future reactor designs and specifically consider the Gen-IV situation and new small modular reactor (SMR) concepts. One deals with nuclear waste management. Technological issues are then addressed and safety issues are discussed pointing the need of technology support organisations, advances and challenges in digitalisation and in novel design tools. The importance of nuclear research, facilities and infrastructures is underlined. The question of education and training of manpower is examined. One objective is that of attracting young graduates from higher education in the nuclear industry. Another is the training of employees in the utilities to give them the proper scientific background and the culture of safety management. Two chapters deal with engineering issues including that of managing nuclear projects, questions of handling safety requirements while controlling costs and complexity. The pertinence of international support of nuclear projects in emerging countries is examined.

Societal issues are considered in the last fourchapters. An assessment is provided of the impact of global nuclear activities on human health during the last fifty years. Since safety is a central issue in the operation of NPPs, it is necessary to the perception of risk to be analysed compared with the real hazards. The report underlines the need to improve public awareness and considers the governance needed for that purpose and the organisation and roles of the different stakeholders to improve public understanding and limit the risks of increasing regulations that would complicate the development and operation of nuclear facilities.